ISO 14001 REQUIREMENTS

61 REQUIREMENTS CHECKLIST AND COMPLIANCE GUIDE

JACK KANHOLM

AQA Co. ▪ Los Angeles

ISBN 1-882711-18-1

AQA Co.

AQA Co.
155 S. El Molino Av.
Pasadena, CA 91101
Phn: (626) 796 9000
Fax: (626) 796 9070

Printed in the United States of America

To my wife, Pia

PREFACE

This book identifies and explains 61 distinct requirements contained in the ISO 14001 standard Section 4, Environmental Management System Requirements. This section of the standard is the specification for the certification of environmental management systems. The requirements are interpreted and explained for the purpose of a certification audit.

The purpose of this book is to identify all auditable requirements and explain how these requirements should be satisfied on the level of documentation, implementation and records; and to offer some insight on how auditors will verify conformance to each requirement. The strength of this book is a systematic and disciplined approach to the identification of all auditable requirements, and reinterpretation of these requirements into a definitive and concrete action list that must be implemented to achieve full conformance. This first step of precisely identifying the requirements and interpreting them correctly is the most difficult one for most organizations. Once this step is taken, the project of implementing ISO 14001 can be properly planned and be carried out with confidence that the effort will result in successful certification.

This book is intended for managers and teams charged with the responsibility to implement and get certified ISO 14001 environmental management systems in their organizations. While the book is written specifically for this group, anyone interested in detailed and comprehensive interpretation of ISO 14001 should find it very interesting. The book could be successfully used for training auditors, and would provide valuable reference for registrars, regulators and consultants.

The book is the leading title in the AQA ISO 14001 series. The other three books are: *ISO 14001 Documentation* and *ISO 14001 In Our Company*, both written by Jack Kanholm; and *ISO 14001 and the Law*, written by S. Wayne Rosenbaum. The role of this book in the series is to introduce the ISO 14001 requirements and explain the basic scope of the actions that must be taken to achieve conformance. The other books contain more specialized information regarding documentation, training and the legal aspects. The series is annotated on the last two pages of this book.

The author would like to express his deep gratitude to Ernani Pires, John Brookes and Greg Hansa of SGS International Certification Services. Their support made this book possible. Much of the material for this book emerged from long discussions between the author and John Brookes, whose pioneering work with ISO 14001 was an important source of information and a great inspiration. Greg Hansa generously shared his practical experience in conducting on-site certification audits.

GOOD LUCK WITH YOUR ISO 14001 IMPLEMENTATION PROJECT AND CERTIFICATION OF YOUR ENVIRONMENTAL MANAGEMENT SYSTEM !

Jack Kanholm

REQUIREMENTS

4.1. GENERAL REQUIREMENTS

4.2. ENVIRONMENTAL POLICY

4.3. PLANNING

4.3.1 Environmental aspects

4.4.2 Training, awareness and competence

4.4.3 Communication

4.4.4 Environmental management system documentation

4.4.5 Document control

3.0 DEFINITIONS

ISO 14001 Section 3, *Definitions,* contains important clues for defining the scope of application of the standard and for interpreting specific requirements stipulated in Section 4. While some definitions simply explain a common sense meaning of the terms used, many definitions include statements that reveal important principles and intentions of the standard.

In the following, all 13 definitions included in Section 3 are quoted (in italics) and are explained with regard to their impact on the interpretation of the standard.

3.1 Continual Improvement

ISO 14001 DEFINITION

"Process of enhancing the environmental management system to achieve improvements in overall environmental performance in line with the organization's environmental policy. NOTE: The process need not take place in all areas of activity simultaneously."

COMMENTS

It is interesting to note that the definition of continual improvement focuses the improvement process on enhancement of the environmental management system rather than the environmental performance itself. Improvement of environmental performance is the intended outcome, but not a strict requirement. To be sure, the environmental performance must show improvement to meet the stated objectives and targets, including compliance with regulatory requirements. However, once the objectives are met there is no obligation to set new, higher standards in all attained areas.

On the other hand, the environmental management system must be continually improved to be ever more efficient in ensuring that the environmental performance is at least maintained on the level consistent with the environmental policy.

The role and scope of the continual improvement element in ISO 14001 is somewhat ambiguous. There are two problems. One is the confusion whether it requires improvement of the management system or environmental performance, and the other is how continual improvement should be planned, implemented and verified. The requirement to include the commitment to continual improvement in the environmental policy elevates this element to the status of great importance but, other than the policy commitment, there are no further requirements for planning, implementing and checking continual improvement programs. The issue is simply dropped. As a result, continual improvement, other than a policy statement, will be very difficult to audit and enforce.

3.2 Environment

ISO 14001 DEFINITION

"Surroundings in which an organization operates, including air, water, land, natural resources, flora, fauna, humans, and their interrelation. NOTE: Surroundings in this context extend from within an organization to the global system."

COMMENTS

This definition is central to understanding the scope of application of the standard. There are two issues that emerge from this definition: conservation of natural resources and the workplace environment.

Regarding the first, note that in addition to the physical environment — air, water, land, flora, fauna, and humans — the natural resources are also included. Consequently, energy and raw material use must be considered, even when their extraction or production does not impact the physical environment.

The second issue is more complicated. Inclusion of *humans* in the definition, and the statement that surroundings include areas within the organization, suggest that the work environment should also be included in the scope. However, the introduction to the standard contradicts this interpretation, stating clearly that the standard is not intended to address occupational health and safety issues. There is still no consensus on how to interpret this contradiction. A firm interpretation of where the environment stops and occupational health begins will emerge with time. For now, auditors include the workplace environment in the scope of the audit to the extent the audited organization itself makes it a part of its environmental management system, i.e., where the environmental policy, objectives, or procedures address the occupational health and safety issues. However, auditors should not object to the exclusion of health and safety from the environmental management system, especially where there is an independent system (procedures and training programs) addressing these issues.

3.3 Environmental aspect

ISO 14001 DEFINITION

"Element of an organization's activities, products or services which can interact with the environment. NOTE: A significant environmental aspect is an environmental aspect which has or can have a significant environmental impact."

COMMENTS

The concept of an environmental aspect is fundamental to the ISO 14001 approach. If an organization does not correctly identify its relevant environmental aspects, it will not be able to successfully implement the standard.

The definition is very broad. It encourages a broad review of activities, products and services, resulting in identification of a large number of aspects. Note that an aspect is defined in terms of its *interaction* with the environment rather than its *impact* on the environment. *Interaction* is neutral — it may be positive or negative — and is much broader than *impact*. The word *can* implies that potential interaction with the environment also must be considered.

Significance of an aspect should not be taken into account at the stage of identifying relevant aspects. The list of identified environmental aspects is meant to be a broad pool from which significant aspects will be selected later.

The Note defines a significant impact without defining what *significant* means. This is intentional and consistent with the ISO 14001 methodology. Organizations implementing the standard must themselves define the meaning of *significant*, in the context of their products, services, location, views of interested parties, environmental regulations, and other such considerations.

3.4 Environmental impact

ISO 14001 DEFINITION

"Any change to the environment, whether adverse or beneficial, wholly or partially resulting from an organization's activities, products or services."

COMMENTS

The definition of environmental impact is also very broad. *Any* change to the environment is an impact. Every identified environmental aspect should be reviewed for present and potential impacts. Not all aspects can be associated with specific impacts, but most can. Again, at this stage the importance of an impact should not be considered. Association of environmental aspects and their corresponding impacts is an important first step in determining the significance of an aspect.

3.5 Environmental management system

ISO 14001 DEFINITION

"That part of the overall management system which includes organizational structure, planning activities, responsibilities, practices, procedures,

processes and resources for developing, implementing, achieving, reviewing and maintaining the environmental policy."

COMMENTS

The definition is strait forward and self-explanatory. The interesting part is the role of the environmental policy. The policy is the driver for the whole environmental management system, and in fact, its raison d'être. Consequently, the suitability, effectiveness, and even conformance of an environmental management system is assessed against its ability to achieve the environmental policy goals. Environmental system auditors use the policy as one of the principal audit criteria.

3.6 Environmental management system audit

ISO 14001 DEFINITION

"Systematic and documented verification process to objectively obtain and evaluate evidence to determine whether an organization's environmental management system conforms to the environmental management system audit criteria set by the organization, and communication of the results of this process to management."

COMMENTS

For those experienced with quality management system auditing (ISO 9000, for example) the definition is strait forward and does not merit any comments. However, without such experience this definition does not help to visualize what a system audit is.

An environmental management system audit is carried out by a team of internal, second-party or third-party auditors. Internal auditors are selected from within the company. They do not have to be certified professional auditors but should have some documented auditor training. Second-party audits are conducted by an organization having direct relationship with the auditee — a customer, for example. Third-party audits are conducted by independent registrars accredited by a governmental or other official body to issue certificates of conformance with the standard. Third-party auditors are formally qualified professionals.

Preparing for an internal audit, the audit team reviews the environmental policy, the system manual and procedures, and other documents defining the requirements of the environmental management system. Based on the requirements of the system and the ISO 14001 standard, auditors prepare checklists with issues and questions for the audit.

The audit process is basically a systematic verification that policies, procedures and instructions comprising the management system are being followed throughout the organization. In all relevant areas, departments and functions of the company auditors interview personnel; review documents, logs and other records; check relevant equipment; verify that personnel are competent and adequately trained; test communication systems; review environmental performance data; and assure themselves that the management system is capable of identifying problems and responding with corrective actions. For an average mid-size company, six to 10 auditor days are required to complete the audit. Internal audits are often distributed throughout the whole audit cycle (typically a year) with partial audits of individual departments, activities or elements of the system.

Inadequate implementation, violation of procedural requirements, and other deviations from planned arrangements and requirements of the ISO 14001 standard are documented as Nonconformances and/or Corrective Action Requests (CARs). A follow-up audit is required to verify that the requested corrective actions are implemented and that they are effective.

An environmental management system audit is not, and should not comprise, a compliance audit. While verifying how the the organization implements its commitment to comply with laws and regulations auditors will review a sample of pertinent data, and may thus incidentally identify a noncompliance. However, they will not engage in a deliberate and systematic verification of compliance. The focus of the audit is the implementation and effectiveness of the environmental management system.

3.7 Environmental objective

ISO 14001 DEFINITION

"Overall environmental goal, arising from the environmental policy, that an organization sets itself to achieve, and which is quantified where practicable."

COMMENTS

Environmental objectives translate the general commitments of the environmental policy into a number of specific goals that need to be achieved in order to implement and fulfill the policy. For example, when the environmental policy states a general commitment to reduce the consumption of natural resources, the corresponding objectives may be to reduce waste and consumption of materials; reduce energy required in manufacturing; and reduce vehicle miles. In particular, specific objectives should be established to fulfill the policy commitments; where there has been, or will be a difficulty in meeting regulatory or other requirements; and where there is lack of consistent, reliable control in any area.

Although not stated in this definition, there should be correspondence between environmental objectives and significant aspects. This does not mean that all significant aspects must be translated into specific objectives, but all should be considered. When there is no objective corresponding to an established significant aspect, there should be a valid reason. It may be that the aspect is fully controlled and further improvement is not possible, or that it is not prioritized at this time.

The definition states that the environmental objectives should be quantified where practicable. This requirement applies also to targets (see Definition 3.10). While objectives may be more general and do not necessarily have to be quantified, the targets always must be. The purpose of a target is to measure progress toward achieving the objective, which cannot really be done without this being quantified.

3.8 Environmental performance

ISO 14001 DEFINITION

"Measurable results of the environmental management system, related to an organization's control of its environmental aspects, based on its environmental policy, objectives and targets."

COMMENTS

The definition is quite strait forward and intuitive. It highlights the need to express environmental objectives, targets and significant aspects in terms that can be measured.

3.9 Environmental policy

ISO 14001 DEFINITION

"Statement by the organization of its intentions and principles in relation to its overall environmental performance which provides a framework for action and for the setting of its environmental objectives and targets."

COMMENTS

Environmental policy in ISO 14001 has a special role. When further developed into objectives and targets, it becomes in fact a specification for the desired environmental outcome or even specific performance. The whole management system is designed and implemented to achieve the policy, and the effectiveness and suitability of the system is assessed on the basis of its ability to fulfill this mission.

This approach is very different from ISO 9000. In a quality management system, the quality specification is written into a contract or is established with product design. A quality policy is expected only to express a general commitment to quality and does not need to be further developed into objectives and targets.

An important part of this definition of environmental policy is the requirement that the policy provide a framework for action and for setting environmental objectives and targets. There should be a link between every objective and the policy. When an objective is not related to any general commitment in the policy, it may be argued that the policy is not appropriate because it does not create a framework for the objective, or that the objective is illegitimate because it is not consistent with the policy.

To ensure completeness, the environmental policy should be established only after significant environmental aspects of activities, products and services are identified, so that commitments to control all significant aspects can be included in the policy. If a new environmental aspect develops due to changes in existing products, processes, technology or capacity, and the aspect is not supported by the current environmental policy, the policy should be revised. In ISO 14001 the policy is not for ever. It should be regularly reviewed and be revised when necessary — this is explicitly required in Section 4.5, *Management review.*

In theory, the ISO 14001 management system is driven only by the environmental policy and is thus not capable of addressing issues not supported by the policy. In practice this statement is certainly too sweeping, but the theory helps to understand why ISO 14001 dedicates a whole section exclusively to the environmental policy.

3.10 Environmental target

ISO 14001 DEFINITION

"Detailed performance requirement, quantified where practicable, applicable to the organization or parts thereof, that arises from the environmental objectives and that needs to be set and met in order to achieve those objectives."

COMMENTS

An environmental target is a specific milestone in the process of achieving an environmental objective. For example, when the objective is to reduce vehicle miles, there may be a target to reduce personnel commuting miles per person by 10 percent in the first year and five percent in the following year (by carpooling, for example); and a target to reduce transportation miles per manufactured unit by five percent per year for the following five years.

In some cases objectives may be stated in sufficiently precise and quantifiable manners so that further breakdown into targets is unnecessary.

Whenever possible, targets must be stated in terms that are measurable and quantifiable. If progress toward achieving a target cannot be measured objectively, the target has no real meaning. An effort should be made to develop indicators even for intangible and abstract targets, such as improvement of community relations.

3.11 Interested party

ISO 14001 DEFINITION

"Individual or group concerned with or affected by the environmental performance of an organization."

COMMENTS

It is intuitively clear in most cases who is *affected* by environmental performance of an organization, and that interests of affected individuals and groups should be considered. However, who exactly should be *concerned* with an organization's environmental performance seems to be a political matter. ISO 14001 auditors will not engage in the debate. They will expect the organization to have a system for communicating with the public and to consider, in balance with its financial, operational and business requirements, the views and interests of at least those individuals or groups directly affected by its environmental performance.

3.12 Organization

ISO 14001 DEFINITION

"Company, corporation, firm, enterprise or institution, or part or combination thereof, whether incorporated or not, public or private, that has its own functions and administration. NOTE: For organizations with more than one operating unit, a single operating unit may be defined as an organization."

COMMENTS

An interesting element of this definition is the NOTE. A relevant question is: Can an operational unit be considered as an organization when the management of the unit does not have the full independent authority to formulate a complete environmental policy and secure adequate resources to achieve it? The answer seems to be "No". Accreditation and registrar bodies will decide where to draw the line.

3.13 Prevention of pollution

ISO 14001 DEFINITION

"Use of processes, practices, materials or products that avoid, reduce or control pollution, which may include recycling, treatment, process change, control mechanisms, efficient use of resources and material substitution. NOTE: The potential benefits of prevention of pollution include the reduction of adverse environmental impacts, improved efficiency and reduced costs."

COMMENTS

The definition provides valuable information on how to implement the ISO 14001 requirement that an organization commit itself in its environmental policy to prevention of pollution. To satisfy this requirement, the environmental objectives and targets and the programs for achieving them should include specific commitments to employ relevant methods for prevention of pollution listed in the definition.

4.1 GENERAL REQUIREMENTS

The purpose of Section 4.1 is to define the extent of the mandatory part of the ISO 14001 specification for contractual or certification purposes. This mandatory part is the whole of Section 4, *Environmental management system requirements.*

The clarification of which part of the standard provides the mandatory specification was necessary, to establish clear criteria for environmental system audits, especially third-party certification audits. It clears away the confusion whether Annex A, *Guidance on the use of the specification,* can be used as audit criteria, and whether organizations may be excused from any requirement of Section 4 because, for example, a requirement seems not to apply to the organization.

In its short statement, Section 4.1 provides the answer to both questions: Annex A is strictly informative and may not be used as audit criteria; and the whole of Section 4, i.e., each and every requirement, is mandatory, irrespective of whether it is deemed to be relevant or not. In other words, no picking and choosing.

In fact, the most important prerequisite for passing the certification audit is completeness of the environmental management system, i.e., a system that addresses each and every requirement. Auditors may go along with some elements being addressed lightly, or may raise a minor nonconformance; but when a requirement is not addressed at all, a major nonconformance — and thus a failed audit — is automatic. For example, inadequate records of management reviews may be classified as a minor nonconformance, but a complete lack of such records will always be a major nonconformance.

4.2 ENVIRONMENTAL POLICY

The environmental policy drives the entire environmental management system. It communicates the organization's mission, vision, core values and beliefs with respect to the environment, and contains commitments to improve, maintain or control environmental performance with respect to specific significant environmental impacts of the organization's activities, products and services. The mission, vision, values and commitments expressed in the policy should be sufficiently specific to provide a framework for establishing objectives and targets that must be met to implement the policy. Once the policy, objectives and targets are established, the remaining elements of the environmental system are designed and implemented for the sole purpose of fulfilling the policy. The environmental policy is thus a supreme specification for the mission and the purpose of the whole environmental management system.

To fulfill its purpose, the policy cannot be a public relation banner of well-sounding but meaningless phrases. This is different than ISO 9000. While in quality management systems the required quality level is often specified by the customer or is included in design documents, the environmental management system must define its own goals and state them in the environmental policy.

The environmental policy must be formulated and/or endorsed by top executive management. Policies that are signed by a mid-level management representative are suspect. Top-level endorsement of the policy not only signifies commitment to its goals, but also a commitment to provide adequate resources to establish and maintain an environmental management system for achieving these goals.

Everything in the environmental policy is auditable. The policy will be checked against relevant environmental aspects and impacts, legislative and regulatory requirements, and the objectives and targets. There must be a clear correlation between these elements. One of the most important aspects of the audit is to verify that the environmental policy is appropriate and comprehensive. Nonconformances related to the environmental policy are usually classified as major and result in a failed audit.

There may be some legal issues associated with the environmental policy. Legal aspects of implementing ISO 14001 are discussed in another AQA publication, *ISO 14000 and the Law* by S. Wayne Rosenbaum.

4.2 Environmental Policy

REQUIREMENT

Formulate an environmental policy appropriate to the organization's environmental impacts of its activities, products and services.

Every organization's interaction with, and impact on the environment is unique. It depends on the nature and scale of the operations (activities, products and services); history of environmental performance and incidents; location; land use in the immediate neighborhood; proximity to environmentally sensitive natural features (watercourses, forests, habitats, etc.); environmental impacts from other sources in the vicinity; applicable environmental legislation and regulations and the level of enforcement; community and wider political climate; customer and market attitudes; and so forth. The above list is not exhaustive, but it indicates how many unrelated factors can contribute to the environmental position of an organization.

For the environmental policy to be appropriate and comprehensive, management responsible for formulating the policy must be fully aware of all relevant environmental issues and their relative importance. An organization seeking ISO 14001 registration should be able to demonstrate that it has a disciplined and systematic process for identifying and evaluating these issues.

Unlike EMAS and some other European regulations, ISO 14001 does not require a formal initial environmental review and a formal statement of current position. The requirement for a formal review was dropped from ISO 14001 because of legal issues in the United States. Understandably, in the present legal and regulatory regime many companies are not keen on researching and documenting their past environmental practices and performance. Nonetheless, ISO 14001 Appendix A and ISO 14004 — noncompulsory guides for implementing the standard — strongly recommend initial environmental reviews.

While a formal initial environmental review is not explicitly required, organizations must still demonstrate that their management has comprehensive knowledge of the relevant environmental impacts, and thus is able to formulate an environmental policy that is appropriate to the organization. The burden of proof is on the auditee. Auditors cannot be expected to independently verify that the policy is appropriate. Their responsibility is to assure themselves that a suitable process was used to gather and communicate information required for formulating a relevant environmental policy.

Another contentious issue is product life cycle analysis. While ISO 14001 does not require formal product life cycle analysis, it is obvious that in order to identify all relevant environmental aspects related to an organization's products and services, consideration must be given to such issues as impacts associated with making the materials and components used in the product, impacts of the products themselves during their service life, disposal and other end-of-life impacts, and so forth. Product life cycle considerations are especially relevant for those organizations responsible for designing their products and services. When appropriate, auditors will ask for objective evidence that product life cycle was considered in formulating the environmental policy.

DOCUMENTATION

The environmental policy can be either established as a separate document or be included in the environmental system manual. There is no need to have procedures for carrying out the initial environmental review or formulating the environmental policy. These activities take place before the whole management system and its procedures are established. However, there should be a process for updating and revising the policy. This is usually done in the framework of management reviews and can be documented in the Management Review procedure (see Section 4.6). Procedures instructing how to identify environmental aspects and how to establish a list of applicable legal and regulatory requirements (see Sections 4.3.1 and 4.3.2) are also relevant to defining the process of collecting information about the organization's environmental position.

EVIDENCE/RECORDS

The evidence demonstrating that a suitable process was used to ensure that the environmental policy is appropriate and comprehensive includes a list of environmental aspects and impacts; a list of applicable legislation and regulations; and a current environmental review report, if such is established (not mandatory).

AUDIT

The whole audit team will usually participate in the audit of the environmental policy. Auditors will examine the methods used to gather information regarding the current environmental position of the company and review the resulting report (if established). They will also review the list of environmental aspects and the list of applicable legislation and regulations, and will verify that all important aspects and impacts are addressed in the environmental policy. Verification that the environmental policy is appropriate to the organization and its environmental position is the most important element of the whole audit. There can be no meaningful ISO 14001 environmental management system without the environmental policy focusing the system on relevant issues.

To summarize, the requirement that the environmental policy be appropriate to the organization's environmental impacts is not only auditable, but will be audited thoroughly. A nonconformance with this requirement will always be classified as major and will delay the audited organization from receiving ISO 14001 certification.

REQUIREMENT

Ensure that the environmental policy provides a framework for setting environmental objectives and targets.

Establishment of environmental objectives is required in ISO 14001 Section 4.3.3. Environmental objectives are specific goals that need to be achieved in order to fulfill the commitments of the environmental policy.

This requirement that the policy provides a framework for objectives is consistent with the role of the policy. The environmental policy is the driver for the whole environmental management system. Every objective, target, program, and activity of the environmental system must arise from the policy. Legitimacy of objectives is also an issue. One of the reasons for this requirement is that the top executive management who formulated the policy is presumed committed to providing support and resources for those objectives and activities that fulfill the policy. Such support cannot be presumed for objectives that are not supported by the policy.

This requirement further reinforces the underlying principle that the environmental policy in an ISO 14001 system must be relevant, specific, and comprehensive. Meaningless generalities will not be accepted.

DOCUMENTATION

The principle that the environmental policy must provide a framework for objectives and targets should be communicated somewhere in the system documentation. For example, the introductory section of the environmental system manual can explain the importance of the environmental policy, including its role as a framework for objectives and targets. Training materials for personnel should also communicate this role.

In addition, the procedure instructing how to establish and document environmental objectives and targets (see Section 4.3.3) should require that any new proposed objectives be checked for consistency with the environmental policy. If the policy does not provide a framework for the new objective, the policy may need to be amended before the objective is adopted.

EVIDENCE/RECORDS Comparison of established objectives and targets against the environmental policy provides the evidence of conformance. There is no need for any special records.

AUDIT All members of the audit team will usually participate in auditing the environmental policy, objectives and targets. One of the elements of the audit will be systematic comparison of established objectives against the policy. For each objective there should be a corresponding commitment in the policy that generally supports the objective. The correspondence does not need to be one-on-one. A single policy commitment will often provide framework for several related objectives. The audit will also verify that the opposite is true, i.e., that, when appropriate, every commitment in the policy is translated into specific objectives (see Section 4.3.3).

REQUIREMENT ***Include in the environmental policy a commitment to continual improvement.***

The definition of continual improvement in ISO 14001 Section 3.1 defines the improvement process as enhancement of the environmental management system rather than the environmental performance itself. In all, the position of continual improvement in ISO 14001 is somewhat ambiguous. It is only required on the level of policy commitment and management review. There are no requirements for planning, implementing and checking continual improvement programs.

The intent of the standard is to propose a system for achieving and maintaining certain environmental performance levels, but not necessarily perpetually improving performance in all areas. Especially with respect to the regulatory requirements, once the requirements are met there is no obligation to exceed them. On the other hand, the environmental management system must be improved perpetually to be ever more efficient in maintaining and potentially improving the environmental performance. (Continual improvement is also discussed in Section 3, Definitions, and Section 4.6, Management review.)

To conform with this requirement, the environmental policy must contain a statement committing the organization to continual improvement. It would be prudent to define what is ment by continual improvement, i.e., whether it pertains to the environmental management system, environmental performance, or both. An example of a statement that would satisfy this requirement is: "ABC Inc. will strive to continually enhance and improve its environmental management system to ensure that our environmental performance is consistent with applicable legislation and regulations and the environmental goals and principles stated in this policy. Performance of the environmental management system will be monitored through internal and external audits, and the system will be periodically reviewed by the executive management."

DOCUMENTATION

At a minimum, two documents of the environmental management system should reference continual improvement: The environmental policy, stating the commitment to continually improve the management system; and the procedure dealing with management reviews, providing instructions for how opportunities for improvement should be identified and implemented.

EVIDENCE/RECORDS

The principal evidence of implementation are records of management reviews. The records should demonstrate that continual improvement of the environmental system is considered by the management review and that specific improvement recommendations are made. There should also be evidence that the recommendations are implemented. Secondary evidence of conformance are records of corrective and preventive actions aiming at improvement of the management system. Environmental performance data and trends will provide evidence that improvement of the system also results in tangible improvement of the environmental performance.

AUDIT

Verifying conformance with this requirement, auditors will first check the environmental policy for a commitment to continual improvement. Beyond the policy it will be difficult to prove nonconformance, unless the auditee can show absolutely nothing to demonstrate that continual improvement is reviewed and enhancements are implemented. Records that will provide auditors with relevant evidence are management review records, corrective and preventive action records, procedure revision records, and environmental performance data.

REQUIREMENT

Include in the environmental policy a commitment to prevention of pollution.

In principle, ISO 14001 leaves it to the organization implementing the standard to determine what are the important environmental goals that will drive the environmental management system. But there are two exceptions: prevention of pollution and compliance with legal and regulatory requirements. They must always be included in the policy.

Policy commitment to prevention of pollution should be reflected in the planning, implementation, verification, and management review elements of the system. All relevant types and sources of pollution should be identified, and the important and controllable ones should be classified as significant aspects. As required generally for significant aspects, these should be reflected in the environmental policy, objectives and targets, and there should be implemented programs for achieving the objectives. Progress toward reduction or elimination of pollution should be monitored and reported to management for review. Once pollution-related environmental aspects are classified as significant, the system will automatically elevate prevention of pollution to the required level.

The commitment to prevention of pollution does not need to be unconditional. Technical and economical viability and cost effectiveness considerations can be used to set limits on how much resources will be allocated to prevention of pollution. The principle of balancing environmental protection needs with financial, operational and business requirements of organizations is clearly stated in the ISO 14001 standard.

An example of a statement that would satisfy this requirement is: "ABC Inc. will strive to identify processes, materials and products that cause, or may cause pollution, and will implement measures to avoid, reduce or control pollution where technically and economically viable."

DOCUMENTATION

At a minimum, implementation of this requirement should be documented in the environmental policy and two procedures. There should be a policy statement of commitment to prevention of pollution (see example above); a procedure instructing how to deal with pollution-related issues when identifying and classifying environmental aspects; and a procedure requiring that programs for prevention of pollution and their results be evaluated by the management review. The procedures need not be independent. Instructions relevant to prevention of pollution can be included in procedures dealing with environmental aspects and management reviews.

EVIDENCE/RECORDS

The evidence of conformance with this requirement consists of the environmental policy; lists of significant environmental aspects, objectives and targets; documents and records of relevant pollution prevention programs; environmental performance records pertaining to pollution; and records of management reviews.

AUDIT

Assessing conformance, auditors will investigate how pollution-related aspects were identified. They will expect some sort of initial environmental review including life cycle considerations, review of manufacturing processes, status of recycling and treatment programs, etc. The method used must ensure that all relevant pollution issues are identified.

Auditors will verify that the environmental policy includes a clear commitment to prevention of pollution, and that this commitment is reflected in objectives and targets. They will also review programs aimed at prevention of pollution and the resulting performance, and will investigate how management reviews address pollution-related issues.

In the field, auditors will look for signs of existing and potential pollution hazards caused by inappropriate dumping and storage, leaks, lack of containment arrangements, malfunctioning and/or inadequate filtering and treatment equipment, and other such hazardous installations, processes and practices.

REQUIREMENT

Include in the environmental policy a commitment to comply with environmental regulations and other requirements to which the organization subscribes.

Legislative and regulatory compliance is the most sensitive issue in ISO 14001. A number of important questions come to mind: Does a third-party certification audit include a regulatory compliance audit? How can a small company with limited resources be expected to fully and completely identify all laws and regulations that may apply to its operations? Can a company that does not currently comply with a regulation be certified to ISO 14001? What is meant by a periodical self-evaluation of compliance? And so forth.

Some answers to these and other such questions will emerge from this book, but will be limited to technical, rather than legal interpretation. For an in-depth discussion of legal issues, please refer to *ISO 14000 and the Law* by S. Wayne Rosenbaum.

To put all compliance-related issues into proper perspective we must remember that ISO 14001 is an environmental *management system* standard and not a standard for environmental performance. It does not establish any absolute requirements for environmental performance and it does not intend to increase or change an organization's legal obligations. These statements are quoted from the introduction to ISO 14001. With respect to legal and regulatory compliance, the standard requires a *commitment* to comply rather than unconditional compliance. Using the standard to enforce regulatory compliance is contrary to its intention and purpose.

The requirement for a *commitment* to comply should be interpreted as a requirement for an organization to take all necessary and reasonable actions to ensure compliance, including identification of applicable laws and regulations, determination of compliance status, implementation of corrective actions in areas where full compliance is not yet achieved, and a system for continually maintaining required performance. From the above considerations it is clear that an organization that does not currently comply with a regulation can be certified to ISO 14001 providing that the organization has identified the area of noncompliance and has implemented programs to improve its performance to achieve compliance. In order to maintain the certificate, the organization must demonstrate continual progress toward achieving compliance, and ultimately achieve it. This interpretation is consistent with the letter and intent of the standard.

In addition legal and regulatory compliance, the standard also requires a commitment to comply with other requirements to which the organization subscribes. This may include industry codes of practice, special agreements with public authorities, voluntary guidelines and standards, contractual requirements, and internal performance criteria.

An example of a statement that would satisfy the requirement for a policy commitment to comply with regulations is: "ABC Inc. is committed to comply with applicable environmental laws, regulations, contractual requirements and other environmental requirements to which ABC Inc. subscribes. To achieve and maintain compliance, ABC Inc. will identify such applicable requirements, will establish objectives and targets for achieving and/or maintaining compliance with the requirements, and will implement measures to monitor compliance with relevant requirements." Everything in this statement is required by ISO 14001 anyway, whether pledged in the policy or not.

DOCUMENTATION

To satisfy this specific requirement the environmental policy must contain a statement of commitment to comply with environmental laws, regulations and other requirements to which the organization subscribes. In addition, a number of procedures will instruct how to identify applicable regulations, establish corresponding objectives and targets, evaluate compliance, and review the system for ensuring compliance. Specific requirements pertaining to these issues will be discussed in subsequent sections.

EVIDENCE/RECORDS

At this stage, the evidence of conformance consists of the commitment statement in the environmental policy. Other requirements pertaining to regulatory compliance will be identified and discussed in subsequent sections.

AUDIT

Assessing conformance, auditors will review the environmental policy to identify a clear commitment to comply with legislative, regulatory and other requirements to which the organization subscribes. Absence of a commitment to comply with the *other* requirements will be noted as a nonconformance with ISO 14001. In fact, auditors will ask what, other than legislative and regulatory requirements is taken into consideration.

Auditors will also review other environmental management system elements pertaining to regulatory compliance. These will be discussed in subsequent sections.

REQUIREMENT

Document and implement the environmental policy and communicate it to all employees.

The environmental policy can be established as a separate document or be included in the environmental system manual. The policy should be signed by an executive manager and be controlled with respect to its revision level, i.e., be dated or identified with a revision number.

A typical policy should be a couple of paragraphs long. One-paragraph format often used to formulate ISO 9000 quality policies is not suitable for an environmental policy. To help organize the policy it is a good idea to divide it into sections identified by headings. For example, sections could be dedicated to continual improvement, prevention of pollution, and compliance with legislative and regulatory requirements.

The requirement to communicate the environmental policy to all employees is somewhat redundant with more stringent requirements in Section 4.4.2 for training employees to understand the policy. The word *communicate* should be understood to mean *explain*. All employees should know the most important principles and commitments stated in the policy, and should be able to explain their roles and responsibilities in achieving conformance with the environmental policy. It is a good idea to post the policy at conspicuous locations throughout the company so that all employees can familiarize themselves with it.

DOCUMENTATION

A written environmental policy is in itself sufficient to conform with the requirement to document the policy. With respect to communicating the policy to employees, the environmental manual should explain how the policy is communicated, and there should be documented training programs to explain the policy.

EVIDENCE/RECORDS

The evidence of conformance consists of the written policy itself and records demonstrating that all personnel were trained in understanding the policy.

AUDIT

Auditors will verify that the environmental policy is documented and that it is controlled in accordance with applicable document control requirements. Assessing conformance with the communication requirement, auditors will be interested in how comprehensive was the distribution of the environmental policy and will interview personnel asking about principles and commitments in the policy.

REQUIREMENT

Ensure that the environmental policy is available to the public.

The requirement to make the environmental policy available to the public supports the general principle of communication with interested parties. It should not be construed as an intent to introduce public accountability for failing to implement the policy. If this was so, there would also be a requirement to publish environmental performance data or have the policy independently verified (as is the case in the European EMAS). Organizations should not look at this requirement as a potential liability issue and try to reduce their exposure by formulating vague and meaningless policies. To the contrary, they should see an opportunity in communicating to the outside world their specific environmental goals. The public is more impressed with a specific recycling program or a goal to reduce energy consumption than with general talk about environmental responsiveness.

Any practical way of making the environmental policy available to the public is satisfactory. It can be published on the internet or in a newspaper, announced and explained in a press conference, or simply be displayed in the lobby or at the gate. Written requests for copies of the policy should be honored.

DOCUMENTATION

There is no need for any written instruction explaining how this requirement will be implemented. However, it would be natural to include in the environmental system manual, or even in the policy itself, a statement that the policy is available to the public.

EVIDENCE/RECORDS

Any proof that the environmental policy is available to the public is a satisfactory evidence of conformance. It can be a clip from a newspaper, or simply a demonstration that the policy is displayed in locations accessible to the public.

AUDIT

Auditors will ask what measures are implemented to ensure that the environmental policy is available to the public, and will review the supplied evidence. They will also ask how requests for copies of the environmental policy are handled, and whether there have been such requests.

4.3 PLANNING

Establishing of the environmental policy is followed by a phase of planning management programs for fulfilling the policy. The theory of the planning phase is based on the following sequential steps:

1) Environmental aspects are identified to provide a broad pool of relevant issues;
2) All identified aspects are evaluated to select a limited number of important (significant) aspects;
3) Objectives and targets are established to control and/or improve performance with regard to selected significant aspects; and
4) Programs are developed to achieve the objectives and establish the environmental management system.

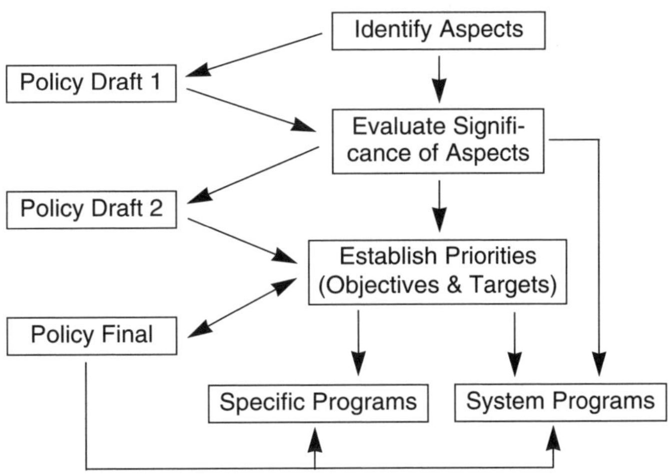

The environmental policy is not directly on this path. It is represented by the significant aspects and by the objectives and targets. The theory is based on the premise that the environmental policy will be fulfilled when the objectives are achieved. Thus, in a technical sense, it is not the environmental policy but the significant aspects and objectives that define the requirements for the environmental management system. A close hierarchical relationship between the environmental policy and objectives is ensured by the requirements that the policy provide a framework for the objectives (Section 4.2.d), and that the objectives be consistent with the policy (Section 4.3.3).

In this sequential approach to planning the environmental system a failure at any step will be carried down. A failure to identify an aspect will preclude it from being evaluated for significance. A failure to properly evaluate significance of aspects will lead to development of irrelevant objectives. A failure to develop an appropriate objective to address a significant aspect will result in lack of a program for achieving control of the aspect. An objective will not be reached if there is no program for achieving it. And, the environmental policy will not be fulfilled if its objectives cannot be achieved.

This is why every step on this path must be perfectly executed and carefully verified. A mistake here is irreversible and may impact the very relevance of the entire environmental management system.

This is theory. But in practice formulation of the environmental policy, identification and evaluation of aspects, and establishment of objectives and targets is an interactive process. All those things are done simultaneously with one influencing the other. The first step is identification of aspects. Then the policy is drafted to set priorities. The priorities are used in evaluating significance of aspects. The policy is then further developed to account for the significant aspects. A set of objectives to fulfill the policy is then proposed. The objectives are reviewed against the list of significant aspects. Adjustments to aspects or objectives are then translated into policy revisions. And so forth. This is the only way to work out a comprehensive, balanced and consistent policy and objectives.

4.3.1 Environmental aspects

REQUIREMENT

Identify environmental aspects of the organization's activities, products and services.

Recalling the definition from Section 3, environmental aspects are elements of an organization's activities, products or services which can interact with the environment. Formal classification of aspects into the three categories (activity, product, service) is not required. However, all three categories should be represented in the list of identified environmental aspects.

Activity-related aspects are specific elements of such activities (or process) as transportation, handling and storage of chemicals; emissions of gases to atmosphere; discharges of waste water; disposal of solid waste; energy consumption; emission of excessive noise or unpleasant odors; operation of underground storage tanks; shut down or start-up of machines or systems; employee transportation; etc. Activity-related aspects that pertain directly to manufacturing processes can also be considered as an independent category or be included together with product-related aspects.

Examples of product-related aspects are: material requirements (kind and quantity); elements of manufacturing processes (often classified as activity-related aspects); material utilization and waste; byproducts; disposal of product packaging; aspects arising from intended and unintended use of products (emissions, noise, energy consumption of products in service); disposal of products at end of life, etc. While complete life cycle analysis is not required, some life cycle consideration should be taken into account.

Service-related aspects are usually more intangible and may include a very wide variety of categories. For example: use of supplies (especially paper); service, maintenance and cleaning of equipment; traffic generated by customers; energy consumption for air conditioning and heating; use of disposable food containers; transportation requirements, means and routing; and so forth.

Precise distinction between aspects related to activities, products and services is not always possible, and is not really important. By naming these three categories, the intent of the standard is to encourage a broad view of an organization's impact on the environment.

Identifying environmental aspects, organizations should consider not only their normal operating conditions but also test, start-up and shut down conditions, and reasonably foreseeable or emergency situations.

At this stage of identifying environmental aspects the importance of the aspect (or rather its impacts) should not be considered. The objective of this exercise is to identify a broad pool of diverse aspects relevant to all facets of operations and products.

The process of identifying environmental aspects should be methodical and deliberate, and be based on solid and comprehensive information. The primary sources of such information are initial environmental reviews, product life cycle considerations, reviews of applicable legislation and regulations, community feedback, and so forth. Although ISO 14001 does not categorically require conducting formal environmental reviews and life cycle analysis, an organized and verifiable method must be used for collecting the baseline information necessary to correctly identify all relevant environmental aspects. A purely intuitive, mental process will not be accepted. (For further discussion of related issues, refer to requirements for the environmental policy.)

The system must be also capable of reviewing the continual relevance of identified aspects and adding new aspects in response to changing circumstances. New aspects may become relevant with product or process changes, procurement of new equipment, construction of new facilities, relocation, changes in the surrounding land use, new legislation and regulations, and so forth. These reviews should be triggered by certain new developments and be included in the periodical management reviews of the system (refer to Section 4.6).

DOCUMENTATION

ISO 14001 clearly states that there must be a procedure for identifying environmental aspects. The procedure should define the scope of and responsibility for conducting preparatory studies to provide baseline information, and outline the methodology and process for identifying and documenting the environmental aspects. Preferably, the process should involve a wide range of management and other personnel with relevant experience and expertise. For example, a special committee could be briefed about the environmental position of the company and then brainstorm a list of aspects. The identified aspects should be documented (recorded), and the procedure should instruct how to establish this record (memorandum or minutes of meetings, for example). The procedure should also require that the process of identifying environmental aspects be applied again in response to changing circumstances, such as product or process changes, expansion, relocation, or external changes in the environment or legal obligations.

EVIDENCE/RECORDS

The most important record is the list of identified environmental aspects. Additional evidence of conformance will be provided by documents and records established in the process, such as baseline reviews and studies, written communication, minutes of meetings, and so forth. For established systems, there should also be evidence that environmental aspects are reviewed in response to changing circumstances.

AUDIT

Auditing the selection of environmental aspects is the first element in the audit of the environmental policy, objectives and targets. All auditors on the team will usually participate. The focus of the audit will be to verify that environmental aspects were identified using a suitable and reliable process, and thus are likely to be appropriate and relevant. Auditors cannot be expected to independently verify the completeness of the list and relevance of each aspect. To be sure, auditors will ask questions about specific aspects or their absence, and will look for obvious omissions. But the degree of confidence in the selection process is the decisive factor. One of the things auditors will be looking for is whether the list includes environmental aspects associated with abnormal operating conditions and emergency situations. For mature systems, auditors will also want to see evidence that identification of aspects is an ongoing process, i.e., that new aspects are identified in response to changing circumstances, and that aspects are regularly reviewed in the framework of managements reviews of the environmental system.

REQUIREMENT

Select those environmental aspects that can be controlled, and over which the organization can be expected to have influence.

ISO 14001 states that organizations need only identify those environmental aspects that they can control and over which they can be expected to have an influence. While this qualification is logical and appropriate, there is a real danger that organizations may interpret their ability to control environmental aspects very narrowly, i.e., exclusively in areas where they have direct and unrestricted decision-making powers. In theory, an organization, especially a subcontractor, could claim that it cannot control its products, because they are designed by its customer; that it cannot control its production technology, because it is dictated by the nature of the product; and that it cannot control or even influence its own suppliers and subcontractors, because they are either specified by the customer or cannot be readily substituted. While some of these limitations may indeed be legitimate in particular cases, organizations should demonstrate a proactive approach in seeking to control or influence the indirect aspects.

The process of evaluating the degree of control and influence that the organization has over an environmental aspect should be deliberate and systematic. There should be some defined criteria and a process of systematically applying these criteria against each aspect. In practice, this evaluation can be merged with the process of analyzing significance of aspects (refer to the next requirement).

DOCUMENTATION

The procedure for identifying environmental aspects should have a section explaining how to evaluate the degree of control and influence that the organization has over its aspects. The procedure may also outline the criteria to be used in the evaluation. Such procedure, or even process is not explicitly required in the ISO 14001 standard. However, it would be very helpful in convincing auditors that decisions to drop certain aspects, because of inability to control or influence them, are not arbitrary.

EVIDENCE/RECORDS

The record of this activity may be a matrix, or a list, that shows whether each aspect can be controlled and/or influenced. There may also be notes or other supporting documents explaining why an aspect is deemed to be beyond control or influence. Again, the standard does not require any such records, but they are invaluable in demonstrating that a systematic and rational evaluation process has been used.

AUDIT

Auditors will often use the issue of control and influence over aspects as a test of the true intentions of the audited organization. When a company automatically assumes that anything outside its in-house operations is beyond its control and influence, auditors take it as a warning that there may be a problem with management commitment to the environmental management system.

Assessing whether the "... aspects over which an organization can be expected to have an influence ..." have been correctly identified will, to a significant degree, depend on the judgment of the auditor. The language in the standard is too indefinite to provide any objective criteria for the assessment. Organizations that want to win the argument over whether an influence can be expected or not should develop procedures and criteria, and establish records of how they evaluate their ability to control and influence their environmental aspects.

REQUIREMENT

Select significant environmental aspects using a disciplined and objective methodology for evaluating importance of environmental impacts.

Selection of significant environmental aspects is the most important step in developing the environmental management system. The selected significant aspects will dictate the environmental policy, objectives and targets, and provide the criteria for the design of the whole environmental management system. The importance of this step cannot be overemphasized.

The list of significant aspects that will emerge from the selection process should be relevant and comprehensive, to ensure that all important aspects are controlled by the environmental system. But the list should not be too long. An environmental system that tries to control a large number of aspects is more likely to fail due to lack of clear purpose and insufficient resources. The best approach is a short but relevant list focusing the environmental system on a limited number of truly important aspects.

The process of selecting significant environmental aspects must use a disciplined approach, i.e., be based on a rational and documented method. The method should include classification of associated impacts and some sort of risk and/or impact analysis — rating the probability of occurrence and consequences.

Aspects can be classified into direct and indirect. Direct aspects are those that can be fully controlled internally, and indirect aspects are those over which the organization can be expected to have influence. Other relevant classification groups are: regulated aspects (legislation and regulations); aspects having pollution-related impacts; aspects related to ecosystems and biodiversity, aspects related to resource utilization; and community-related aspects (amenity). The impacts can be classified into internal, local, regional and global. There can also be a category of impacts that cannot be satisfactorily evaluated due to lack of reliable information or data.

A rudimentary risk analysis can be carried out by rating the probability of occurrence of an impact and its consequence, and then multiplying the probability and consequence ratings to determine the risk factor. The probability can be, for example, rated 1 through 4, with the low rating for likelihood of occurrence only in extreme conditions and the top rating for a likely occurrence in normal operating conditions. The consequence rating can start with 0 (zero) for no measurable impact and go up to 4 for a damaging and irreversible impact.

The overall method for determining significance of an aspect is an exercise of rating risk and other characteristics of an aspect and then combining the ratings in a formula giving the final significance rating. For example, in addition to risk rating, an aspect can be rated with regard to its legal and regulatory relevance, pollution relevance, geographical extent of impacts, relevance to community views and interests, and so forth. The formula for determining the overall significance rating usually involves multiplication or addition of ratings, or a combination of both.

While purely intuitive and mental methods for selecting significant aspects are not acceptable, common sense and intuition should be also used, especially to verify that the results of the systematic approach make sense. The risk analysis and rating systems are only tools aiding in decision making. The final decision is ultimately a judgment.

A special category of aspects are those activities, processes and products that are subject to permitting, reporting or implementation of other special controls required by law or regulations. An effective approach with regard to these aspects is to automatically classify them all as significant, regardless of risk and other factors. ISO 14001 standard strongly implies this approach. An argument can be made that, at a minimum, an organization should consider as significant those aspects that are considered significant by society. Critics may say that this approach reduces ISO 14001 to a regulatory enforcement tool, contrary to the intent of the standard. In theory critics may be right. But in practice there will be many registrars, auditors and regulators out there that will expect all major regulated aspects to be classified as significant.

DOCUMENTATION

ISO 14001 explicitly requires a procedure for selecting significant environmental aspects. The process of selecting significant aspects can be described in the same procedure that deals with the initial identification of environmental aspects (refer to the last two requirements). The procedure should outline methods used for evaluating significance of aspects; assign responsibilities for carrying out the evaluation and for review and approval of the selected aspects; and instruct how to establish a formal, authorized list of significant environmental aspects.

To explain the methods for evaluating significance of aspects, the procedure can include specific instructions on how to rate risks and consequences and calculate the risk factor; define categories of aspects and assign significance ratings for each category; provide the formula for calculating the overall significance rating; and provide charts (matrices) for working out the rating.

EVIDENCE/RECORDS

The principal record is a formal and authorized list of significant environmental aspects. The standard does not explicitly require the list or any specific management level for authorizing the list. However, considering the relationship between the significant aspects and the environmental policy, it is reasonable to expect that both should be authorized at the same executive level. Additional evidence of conformance will be provided by records established in the evaluation process, such as worksheets (matrices) used for risk analysis and significance rating calculation.

AUDIT

Auditors will review the process of selecting significant aspects and the resulting list very carefully. They know the importance of this step.

First, auditors will verify that the significant aspects were selected using a suitable and reliable process. They will review the rating criteria, worksheets and calculations, and ask about the underlying rationale for prioritizing certain categories of aspects.

Then, auditors will examine the actual list of significant aspects. This will be the most important, but also the most difficult, part of the audit. On the one hand, clearly auditors will not have the time and sufficient knowledge of relevant issues to independently verify that the selected list is complete and appropriate. On the other hand, they cannot just passively accept the list only because a disciplined selection method was used. The list is too important. The audit is likely to look more like negotiations than a systematic review of objective evidence. Auditors will ask why certain aspects are not included in the list and the audited organization will explain the reasons and defend its final selection. Auditors will want to broaden the list to be sure that the registrar, accreditation body, EPA, and other interested parties will not reproach them for accepting a system with inadequate policy and objectives. The audited organization will want a narrow list including only truly important aspects that can be effectively controlled. If indeed an organization has selected its significant aspects responsibly, it should decisively defend its selection.

Auditors will pay particular attention to those aspects pertaining to legislative and regulatory requirements and prevention of pollution. Commitments to comply with laws and to prevent pollution are required in the environmental policy. By implication, there should be some related significant aspects to provide focus for these commitments. Absence of such aspects would make the policy feel irrelevant and hollow. Especially with regard to regulated aspects, many auditors will expect that at least those aspects that require permitting or reporting should be classified as significant.

4.3.2 Legal and other requirements

REQUIREMENT

Identify and have access to applicable legislative, regulatory and other requirements.

To comply with this requirement an organization must have a procedure for identifying which environmental laws and regulations apply to its operations and what they require, and for keeping this information up to date. The organization should also keep track of other environmental requirements to which it subscribes, such as industrial guidelines and standards, agreements with public authorities, and contractual requirements.

Information about environmental regulations can be researched from federal, regional and state agencies; counties; cities; industrial and professional associations; or specialized professional services. The result of the research should be a list and a library of all applicable regulations. In some cases it may be appropriate to transcribe some regulatory requirements into internal specifications. If a professional service is contracted, there may be no need to maintain a complete in-house library, providing that the full text of relevant regulations can be quickly obtained from the service. But under any circumstances there should be available in-house a complete list of applicable regulations and a summary of the regulatory requirements.

Identification of applicable laws and regulations may have important legal aspects for an organization. They are discussed in *ISO 14000 and the Law* by S. Wayne Rosenbaum.

DOCUMENTATION

ISO 14001 explicitly requires a procedure for identifying legal and regulatory requirements. The procedure should assign a person (function) responsible for researching information about relevant laws and regulations, and for establishing a list and obtaining copies of applicable regulations. The procedure should also instruct how and how often the list of applicable regulations should be reviewed and updated.

EVIDENCE/RECORDS

The primary evidence of conformance is a list of applicable regulations and copies of these regulations. The list should be released as a controlled document, i.e., it should be authorized and be identified with respect to its revision status. There should also be evidence that the list is periodically reviewed and updated.

AUDIT

Assessing conformance, auditors will review the relevant procedure and will ask for the list of applicable regulations. They will also verify that copies of these regulations are accessible. Maintaining a list and copies of regulations is not explicitly required in ISO 14001, but this is practically the only way to convince auditors that the organization has *identified* and *has access* to all laws and regulations that apply to its operations. An alternative is to retain a professional service that can make copies of the regulations available upon request. In this case, auditors may want to test how quickly a specific regulation can be obtained. A verbal summary of the regulations and a statement that the texts are available in the public library will not be sufficient to demonstrate conformance.

Auditors cannot be expected to independently verify that the list of applicable regulations is complete. Their job is to assess the process of establishing the list rather than verification of the list itself. However, auditors will usually have a good understanding of regulatory requirements and should be able to identify any obvious omissions. An incomplete list is a serious nonconformance, especially if the omissions are carried through the environmental policy, objectives and the program of internal compliance reviews.

Auditors will also verify that the list of applicable regulations is periodically reviewed and updated. There should be some evidence of such review, even when it does not identify any changes in regulatory requirements and the list is not revised. Copies of the regulations should always be the latest revisions.

4.3.3 Objectives and targets

REQUIREMENT

Establish and maintain documented environmental objectives and targets at all relevant functions within the organization.

Environmental objectives are goals that must be set and achieved to fulfill the environmental policy. The implication of this definition is profound. It establishes a premise that a policy commitment that is not translated into a specific objective will not be implemented and fulfilled. Objectives are the link between the policy and the management programs for implementing the policy.

An environmental target is a specific and quantified milestone that must be reached in order to achieve an objective. In some cases objectives and targets may be combined. Simple and specific objectives that are quantified and have a time frame for achieving them may not need to be further developed into targets. Whether targets are developed independently or are contained within an objective is just a question of format. In requiring targets, ISO 14001 communicates the principle that, in the end, the commitments of the environmental policy and the associated objectives be translated into specific plans to achieve certain goals within a specified time frame.

ISO 14001 requires that objectives and targets be quantified where practicable. It means that they must be quantified whenever it is desirable and possible. As it is almost always desirable to quantify a target, and there are known methods to measure almost anything, a target that is not quantified has a good chance of being rejected. Objectives and targets must also specify a time frame. A goal to reduce emissions by 10 percent is meaningless if it does not stipulate whether the reduction will be achieved in one year or three.

Objectives and targets must be established at appropriate functions and levels within the organization. Unlike the environmental policy, which must be formulated by the top executive management, objectives and targets may be specific to a function, department, project, product group, or site; and can be established by lower level managers. For example, a general policy commitment to reduce energy consumption may be translated by the production department into an objective to reduce energy used in manufacturing, by the transportation department into an objective to reduce transportation miles, and by the administrative departments into an objective to reduce the energy used for air-conditioning offices. Failure to develop objectives at a relevant function is a nonconformance. The principle expressed in this requirement is that no relevant function or level within an organization can be exempted from active participation in the effort to implement the environmental policy.

DOCUMENTATION

ISO 14001 does not explicitly require a procedure for establishing and documenting environmental objectives and targets. Nonetheless, auditors will expect that the environmental system documentation explains somewhere the concept and the role of objectives and targets, the responsibility for establishing them, and the manner in which they should be documented. Even though not required, a procedure should be established to explain the system, especially in larger organizations. The procedure could explain what is the role of objectives and targets; how they are derived from the environmental policy and significant aspects; what levels of management are responsible for proposing and/or establishing the objectives; who is responsible for reviewing and approving them; and how objectives and targets should be documented. If there is no dedicated procedure dealing with these issues, the environmental system manual should at least outline the system.

EVIDENCE/RECORDS

The record and evidence of conformance consists of lists, memos, minutes of meetings, or other such documents specifying environmental objectives and targets. ISO 14001 explicitly requires that the objectives and targets be documented. Irrespective of format, these documents must be controlled with regard to their approval status, revision level and distribution. There should also be evidence that objectives and targets are regularly reviewed and, when appropriate, are revised to account for changing circumstances and new developments. Such reviews and changes are usually documented in minutes of management review meetings.

AUDIT

Assessing conformance with this requirement, auditors will note whether objectives and targets are quantified and specify a time frame. A target that is not quantified and/or does not state a time limit will always be questioned. Auditors will accept unquantified targets only in the rare cases when there are no meaningful methods to measure performance. And even in these cases the audited organization will have to explain how they determine progress toward reaching the objective or target.

Auditors will verify that objectives and targets are established at all relevant levels and functions within the organization. For example, if an objective calls for recycling of paper, auditors will verify that targets are established for all departments that use and/or receive considerable amounts of paper.

The audit will also verify that objectives and targets are properly reviewed, approved and documented, and that the documents are controlled.

REQUIREMENT

Ensure that significant environmental aspects are considered in setting environmental objectives and targets.

While in theory ISO 14001 requires the environmental policy to provide a framework for developing environmental objectives, in practice the policy may be too general to lead to specific issues. The list of significant environmental aspects is a better reference for developing the objectives. It is natural to establish a one-on-one correspondence between the aspects and objectives because they both address specific issues at the same level. Consistency between the objectives and the environmental policy will be achieved automatically, as the significant aspects are also considered in formulating the policy.

There is no requirement that objectives be established to address every identified significant aspect, but every significant aspect must be considered in setting the objectives. Whenever a significant aspect is not followed with a corresponding objective, there must be a good reason. It may be because of limited resources and a need to prioritize, or because there is no prospect for meaningful improvement of environmental performance with relation to the aspect.

There are two strategic approaches that can be adopted (both compatible with this requirements of ISO 14001): The first is to set objectives only in relation to those significant aspects where there is a prospect for meaningful improvement or where there is deficiency, for example, noncompliance with regulations. The second approach is to set objectives corresponding to all significant aspects, in which case some objectives will be to maintain performance at current level without any improvement. The advantage of the second approach is that it establishes an unbroken hierarchical order between significant aspects, objectives and management programs (refer to next section), but the idea of objectives that do not set improvement goals may be an alien concept to many people.

The correspondence between significant aspects and objectives does not need to be a direct, one-on-one relationship. In some cases it may be appropriate to lump several aspects together and address them by one objective, and there may be situations where several objectives will be established to address one aspect. For example, excessive use of paper may be addressed by an objective to reduce direct mail and by another objective to reduce internal consumption of paper.

DOCUMENTATION

The general procedure explaining how to establish objectives and targets (see previous requirement) should instruct that all significant environmental aspects must be considered in setting the objectives. If such procedure is not established, the environmental system manual should at least state this principle.

EVIDENCE/RECORDS

Comparison of the list of significant environmental aspects against the list of environmental objectives provides the evidence of conformance. For those aspects that are not addressed by corresponding objectives there should also be some evidence that the aspects were considered and a decision was made to leave them out. Such evidence can be provided by minutes of meetings, reports, memos, or some other record of the decision-making process.

AUDIT

Assessing conformance, auditors will compare the list of environmental objectives against the list of significant environmental aspects. Auditors will ask for explanation in cases where a significant aspect is not addressed by a corresponding objective. It does not mean that there must always be a one-on-one correspondence between significant aspects and objectives. When appropriate, one objective may address several aspects, or an aspect can be left out temporarily without a corresponding objective because of limited resources and need to prioritize. But each significant aspect must be at least *considered* in establishing objectives, and auditors will verify that it was. The objectives will also be compared against the environmental policy (refer to the last requirement in this subsection).

REQUIREMENT

Ensure that legal and other requirements are considered in setting environmental objectives and targets.

This requirement will be satisfied by default when all regulated environmental aspects, i.e., those subjected to legislative and regulatory requirements, are automatically classified as significant, and when all significant aspects are automatically addressed by an objective. This approach is not required by, or even suggested in, the ISO 14001 standard, but it introduces certain hierarchical order in the system, and is therefore recommended throughout this book. Its advantage is that the selection of all issues to be controlled by the environmental management system is done at the stage of selecting significant environmental aspects, and then, all these issues are automatically accounted for in establishing objectives, targets, management programs and monitoring and measurement systems. Establishing objectives for all regulated aspects does not automatically mean that performance in all regulated areas must be continually improved. Where compliance is achieved, the objective may be simply to maintain performance on current level.

At a minimum, objectives must be set in all areas that do not comply with applicable laws and regulations.

The requirement to consider legislation and regulations does not mean that there always must be a specific objective for every piece of applicable regulation. When appropriate, one objective may address a group of related regulations. But the generalization may not be carried too far. The statement "ABC Inc. will comply with applicable regulations" is not an objective, but a policy commitment. In any event, whether objectives are general or specific does not really matter, because they must be further developed into targets; and targets must be specific per definition. In principle, there should be a specific target to meet, maintain or exceed every applicable regulation.

When establishing objectives pertaining to a regulatory requirement, three cases should be considered: 1) Organization does not comply; and the objective is to achieve compliance; 2) Organization complies but does not want to exceed the requirement, and the objective is to maintain current performance; and 3) Organization complies and wants to further improve its performance to exceed the requirement, and the objective is to improve performance to a new, higher level.

For those regulations that impose a specific performance value, or when the performance value can be worked out, the corresponding objectives and target must be also quantified. In the first and second case listed in the paragraph above the targets would have the same performance values as those specified in the regulation.

DOCUMENTATION

The general procedure explaining how to establish objectives and targets (see the first requirement in this subsection) should instruct that all applicable legislative and regulatory requirements and other requirements to which the organization subscribe must be considered in setting the objectives. If such procedure is not established, the environmental system manual should at least state this principle.

Comparison of the list of applicable regulations against the list of environmental objectives and targets provides the evidence of conformance.

EVIDENCE/RECORDS

Assessing conformance, auditors will compare the list of environmental objectives and targets against the list of applicable regulations. In cases where a regulation is not addressed by a corresponding objective, auditors will go further to verify whether there are proper management programs for maintaining compliance, including a system for measuring and monitoring the regulated aspect (required in Section 4.5.1). If there is compliance and a monitoring program is established, auditors will be satisfied.

AUDIT

Auditors will verify that in every case where the organization does not comply with legal and regulatory requirements there are objectives and targets to meet these requirements. If not, auditors will not hesitate to issue a major nonconformance, resulting in a failed audit.

REQUIREMENT

Ensure that views of interested parties are considered in setting environmental objectives and targets.

Implementation of this requirement to consider views of interested parties will differ considerably from one organization to another. It will depend on the nature of the operations and products, location, land use in the immediate neighborhood, and many other factors. For a chemical plant located in proximity to residential areas, or for a large company providing a popular product to the consumer market, consideration of the views of neighbors or customers may be essential. On the other hand, a small service company located in an industrial park may not even be able to identify who the interested parties are, much less what their views may be.

To demonstrate conformance with this requirement, organizations must have a system for communicating with external parties, and a process for considering their views. ISO 14001 Section 4.4.3, *Communication*, explicitly requires such system. An intuitive conclusion that there are no identifiable interested parties is not sufficient. If indeed there are none, the system for communicating with external parties will demonstrate it.

The views of interested parties must be considered, but do not need to be automatically addressed by objectives and targets. The views of interested parties may be balanced against technological options and financial and business requirements of the organization — a flexibility that is not available when considering legal and regulatory requirements.

Documentation

The general procedure explaining how to establish objectives and targets (see the first requirement in this subsection) should instruct that views of interested parties must be considered in setting the objectives. If such procedure is not established, the environmental system manual should at least state this principle. In addition, either the procedure or the manual should assign the authority for deciding whether an external view will be addressed by an objective or not and, when appropriate, may outline the general criteria for making the decision.

Evidence/Records

Comparative review of communications with external parties and the list of environmental objectives and targets provides the evidence of conformance. In cases when known external views are not reflected in the objectives, there should be evidence that the views were considered but were not translated into objectives because of the need to prioritize, or for other reasons.

Audit

Assessing conformance, auditors will review correspondence, minutes of meetings, memos, and other such records of communications with external parties to identify those interests and views that could be relevant in setting objectives and targets. In cases when a relevant view is not addressed by an objective, auditors will verify whether the view was considered and will ask why it was left out. While organizations do not have to react to all external interests and views, there is an implied expectation that they will respond to some. A situation where there are important and relevant concerns of interested parties, but none of them are addressed by environmental objectives, may result in a nonconformance, even when there is evidence that all external views have been considered (evaluated). At the outset, an environmental management system that does not address important aspects concerning external parties may be judged to be inappropriate to the organization and its environmental position.

REQUIREMENT

Ensure that objectives and targets are consistent with the environmental policy, including the commitment to prevention of pollution.

The first part of this requirement reinforces the principle that environmental objectives and targets, and by implication the whole environmental management system, must arise out of the environmental policy. In Section 4.2 this principle was first expressed by a requirement that the environmental policy must provide a framework for setting environmental objectives. Definitions of environmental policy and environmental objectives also reinforce this principle. In practice, conformance with this requirement is ensured automatically when the same set of significant environmental aspects is used in formulating the policy and establishing objectives.

At a minimum, objectives must be set in areas where the goals stated in the environmental policy are not attained. In an ISO 14001 system, objectives and targets are the mechanism by which the environmental policy is fulfilled. A failure to set objectives in those areas will result in a major nonconformance.

The second part of this requirement is a reminder that ISO 14001 requires the organization to commit to prevention of pollution, thereby creating the expectation that at least some objectives and targets will be related to this commitment.

A practical way to ensure conformance is to identify pollution-related issues at the stage of selecting significant environmental aspects. It can be done, for example, by assigning important significance rating to pollution-related aspects, thus ensuring that at least some of them will be classified as significant, and will automatically be addressed by objectives and targets. If no such aspects are selected, there is at least a record demonstrating that a systematic and disciplined evaluation process was used, and thus pollution-related issues were properly considered in setting objectives and targets.

DOCUMENTATION

Consistency between the environmental policy and objectives is best documented in those sections of the environmental system manual dealing with the environmental policy and/or management reviews. The requirement for considering prevention of pollution can be communicated in the general procedure explaining how to establish objectives and targets (see the first requirement in this subsection). If such procedure is not established, the environmental system manual should at least state this principle.

EVIDENCE/RECORDS

There is usually no specific record or other evidence of conformance other than a comparative review of the list of objectives and targets against the environmental policy.

AUDIT

To test consistency between the objectives and targets and the environmental policy, auditors will seek to identify for each objective a policy framework, i.e., a commitment to control and/or improve environmental performance in areas to which the tested objective pertains. The test will also go the other way. For each policy commitment auditors will seek to identify specific objectives and targets. There should be correspondence in both directions.

Verifying that the commitment to prevention of pollution is reflected in objectives and targets, auditors will first review the list of environmental aspects and the process of selecting those significant aspects related to pollution. Having familiarized themselves with the pollution-related issues, auditors will then check whether those issues are addressed by specific objectives and targets. Auditors may disagree in cases where some pollution-related aspects are considered trivial, and therefore are not addressed by objectives. A commitment to prevention of pollution should be implemented whenever there is an opportunity to reduce pollution and the cost is reasonable. On the other hand, if there is a significant number of important pollution-related issues, auditors will not object when the less important issues are left out because of the need to prioritize.

4.3.4 Environmental management programs

REQUIREMENT

Establish management programs for achieving environmental objectives and targets.

Development of programs to achieve environmental objectives and targets is the last step on the path of defining and planning the environmental management system. The programs will provide the means for achieving the objectives and fulfilling the environmental policy.

The programs can be either general system-level, infrastructure-type programs — such as training, document control, or monitoring and measurement systems — or more detailed objective-specific programs. System-level programs are required in ISO 14001 irrespective of the actual stated objectives. They are the operational backbone for the environmental management system. Objective-specific programs are more project oriented. They provide specific methods, means and organization for achieving a particular objective.

When considering programs for achieving an objective, an organization should identify the relevant system-level programs and determine whether, in themselves, these programs are sufficient to achieve the objective. If not, additional objective-specific programs should be developed. System-level programs are sufficient when monitoring, training, and implementation of efficient management practices can alone ensure achievement of an objective. For example, if the objective is to reduce emissions by, say, 10 percent, the system-level operational control programs and monitoring programs required in ISO 14001 Sections 4.4.6 and 4.5.1 may be sufficient to achieve this objective. However, if the objective is a 50 percent reduction, one would expect to find special programs involving product design changes or introduction of new technology. Objective-specific programs must be developed when special arrangements are required for achieving objectives. Examples include carpooling, recycling, product modifications, process changes, or introduction of new technology.

In principle, every established objective and target must be followed up with a program to achieve it. It does not mean that there always must be an independent and dedicated program for every objective. In some instances it may be appropriate to use the same program for achieving several related objectives. For example, a program for monitoring and measuring levels of various substances can be used to achieve several objectives related to control of effluents and emissions. There will also be situations where one objective will require several programs. In a multisite organization, for example, each location may have its own independent program to achieve the same corporate objective.

DOCUMENTATION

ISO 14001 does not require a procedure for establishing and maintaining the programs. Such procedure may not be needed in a small organization with a limited number of objectives, where only one person is responsible for establishing the programs. However, in a larger and more complex organization it would certainly be useful to have some written guidelines and an approval/authorization system to ensure coordinated approach and control over allocation of resources.

If a procedure is written, it could instruct how functional and/or operational units of the organization should identify and document programs to address objectives relevant to their activities; how they should seek approval and authorization of these programs; and how progress toward achieving the objectives should be documented and reported. Whether there is a procedure or not, the system for establishing and maintaining the programs must include some coordinating and reporting functions. Someone should be responsible for ensuring that all objectives have their programs, and for reporting progress to the top management (see Section 4.6, *Management review*).

EVIDENCE/RECORDS

The evidence of conformance usually consists of specifications for the programs, including assignment of responsibilities and designation of means and time frame for achieving the objectives; and records showing their status or progress. The records can be in the form of logs, charts, summaries, memos, minutes of meetings, or other such written reports.

AUDIT

Auditors will first verify that there is a program for achieving every established objective and target. They will do it by going through the list of objectives and asking for each what are the associated programs. It would be helpful for the audit if there was a master list or a matrix associating objectives and targets with the corresponding programs. When auditors find an objective that is without a program, they will write it up as a nonconformance.

Auditors will review the programs to verify that there are designated responsibilities for achieving the objectives and that the programs specify the means and a time frame for achieving them (refer to the next two requirements). The program records will also be reviewed to verify that the programs are active and effective.

REQUIREMENT

Assign responsibility for achieving environmental objectives and targets at each relevant function and level of the organization.

This requirement recognizes the simple truth that nothing will happen without leadership and responsibility. At this point the responsibility in question is for achieving specific objectives and targets, i.e., for implementing and operating individual system-level and objective-specific programs (refer to the preceding requirement). Requirements pertaining to responsibilities for implementing and operating the whole environmental management system are included in Section 4.4.1.

Responsibilities for achieving objectives must be assigned at all relevant functions and levels. The principle expressed in this requirement is that no relevant function or level within an organization can be exempted from the responsibility to achieve objectives. If, for example, the objective is to reduce consumption of energy and there are significant energy uses in production and transportation departments, it is not acceptable for only one of these two departments to be responsible for reducing energy consumption (unless the objective specifically pertains to only one department).

DOCUMENTATION

Companies that want to register their ISO 14001 environmental management systems must document the designated responsibilities for achieving objectives and targets. Even though the text of the standard does not explicitly require documentation, in the minds of third-party auditors it is unimaginable that a designation of responsibility would not be documented. Responsibilities may be documented in organizational charts, in job descriptions, in procedures, or in documents defining specific programs. There can also be a matrix depicting all objective-related programs and associated responsibilities.

EVIDENCE/RECORDS

The records and evidence of conformance are written assignments of responsibilities for achieving objectives and targets.

AUDIT

Assessing conformance, auditors will review the list of objectives and targets and verify whether appropriate functions and personnel are assigned with the responsibility to achieve every stated objective. They will expect the responsibilities to be widely spread throughout the organization. A blanket statement such as "The Environmental Representative is responsible for ensuring that environmental objectives and targets are achieved" is not sufficient, unless the organization is very small and it is reasonable and natural for only one person to manage all objective-related projects. To be sure, it is appropriate and even required to appoint one representative with the ultimate responsibility for the environmental management system. However, in this instance the standard requires designation of specific operational responsibilities.

REQUIREMENT

Specify the means and time frame for achieving environmental objectives and targets.

ISO 14001 requires that management programs for achieving objectives and targets be defined in terms of means and time frame. The requirement for a time frame can be easily implemented by stipulating a due date for achieving every established target. Objectives do not need to have due dates as long as they are further developed into detailed targets. Although the time frame requirement is included in the section dealing with management programs, in most cases it will be more natural to specify due dates together with targets. For example, instead of stating "Reduce water consumption by 20 percent" and defining the due date in the project specification, a better way is to simply state "Reduce water consumption by 20 percent by December, 1999."

The extent to which the means for achieving objectives must be defined will depend on the nature and complexity of the management program for achieving the objective. Large and complex programs will have to be defined in terms of specifications, budgets, schedules, personnel allocations, special equipment requirements, and so forth. At the other extreme, for a program as simple as posting a sign, for example, the means may be defined as an item in a checklist or minutes of a meeting. The format and scope for defining the means will vary, but for each management program it must be sufficient to clearly identify the methods and resources.

Many company-wide programs will require different means in different operations and areas. For example, if the target is to reduce hazardous waste by 20 percent, the program should define how much of the reduction will be achieved through product design changes, process changes, or recycling; and should prescribe specific means for each.

Some categories of objectives and targets can be achieved simply by implementing the environmental management system as required by the ISO 14001 standard itself.

The kinds of objectives in this category are general improvements that will automatically result from better training, instructions, communication, performance monitoring, and other operational controls. For these kinds of objectives the programs and means for achieving them may be simply defined as implementation of relevant elements of the environmental management system.

DOCUMENTATION

Although the standard does not explicitly require this, the means and time frames for achieving targets should be documented. Third-party auditors — and no one else for that matter — would never accept a due date that is not written down. Time frames are best documented together with the targets. The means for simple programs can be documented in a checklist or a matrix. For complex programs, the means should be defined in written specifications and procedures.

EVIDENCE/RECORDS

The records and evidence of conformance are documented due dates for achieving objectives and targets, and program specifications and procedures, or other such documents defining the means for achieving objectives.

AUDIT

Auditors will review the list of environmental objectives and targets and verify that the means and the due dates for achieving each target are defined. When a program for achieving an objective is defined as an element of the general environmental management system, auditors will evaluate whether this is sufficient. For objectives that require special projects, auditors will expect these projects to be individually defined and documented.

REQUIREMENT

Update and amend programs as required to account for new developments and new or modified activities, products or services.

This requirement is a reminder that the whole environmental management system, including environmental policy, objectives, targets, and programs, must be continuously updated to account for new developments and changing circumstances. Reminders to keep the system up to date are included in several sections of the ISO

14001 standard — including this section — but the primary requirements for reviewing and updating the system are stated in Section 4.6, *Management review*.

The kinds of situations contemplated in this requirement include addition of new activities, products or services; expansion or extension of existing activities or products; significant material specification changes (metal to plastic, for example); introduction of new technologies or processes; addition of new plants (locations), and so forth. The environmental management system should also be adopted to include unusual or temporary projects, such as construction of new facilities or plant relocation.

DOCUMENTATION

There is normally no need for a procedure explaining how programs are updated to account for new developments and circumstances. It is sufficient to document in the environmental manual a general commitment to keeping the system up to date. Instructions for reviewing and updating the whole environmental management system, including management programs, are best documented in procedures dealing with management reviews.

EVIDENCE/RECORDS

The evidence of conformance are records of actual program modifications. In companies with relatively new environmental management systems such records may not exist because there have not been any significant changes since implementation of the system.

AUDIT

Assessing conformance, auditors will investigate whether there were any significant changes in operations, processes or products since the initial implementation of the environmental system and, if there were, how existing programs were modified to account for these changes. This part of the audit will usually be carried out in conjunction with requirements pertaining to management reviews.

4.4 IMPLEMENTATION AND OPERATION

ISO 14001 Section 4.4 contains requirements for organizational and operational structures of the environmental management system. The role of these elements is to provide the infrastructure and some basic programs for the system, including operational controls. Specifically, Section 4.4. contains requirements for defining the organization responsible for the environmental system; for providing training programs; for establishing communication systems; for documenting the environmental management system and for controlling the documentation; for controlling relevant operations and activities; and for maintaining emergency preparedness.

Most of the organizational and operational elements in ISO 14001 are equivalent to those in ISO 9000. The requirements for defining organization, providing training, documenting the system, controlling documents, and controlling operations and activities are very similar in both standards. There are in fact only two entirely new requirements in the environmental standard: one for establishing communication systems and one for maintaining emergency preparedness. Those companies that have already implemented ISO 9000 can use their experience, and even share some of the quality system elements (document control, for example), in developing the operational elements and programs for ISO 14001.

In addition to providing the infrastructure for the environmental system, some programs required in this section may be also used for achieving environmental objectives and targets (refer to requirements of Section 4.3.4, *Environmental management programs*).

4.4 Implementation and operation

4.4.1 Structure and responsibility

REQUIREMENT

Define, document and communicate authorities, roles and responsibilities for operating the environmental management system.

This section of ISO 14001 does not state any specific requirements with regard to the organizational structure, other than that it must be documented and communicated. Unlike ISO 9000, it does not require any organizational independence. Personnel verifying environmental performance and the internal auditors of the environmental system do not have to be independent from those responsible for the verified or audited activities. It is not clear why ISO 14001 is not concerned with a potential conflict of interest, especially in the case of internal auditing.

While ISO 14001 does not suggest any particular structure for assigning responsibilities and authorities, it contains some specific requirements with regard to the top (executive) management. It requires that the top management be actively involved with the system. At a minimum, the top management must be responsible for formulating the environmental policy; for appointing a management representative; for providing adequate resources to establish, implement and operate the system; and for periodically reviewing the suitability and effectiveness of the system. These mandated responsibilities should be reflected in organizational charts, specifications of functional responsibilities, and procedures.

The organizational structure may be communicated through distribution of documents defining the organizational responsibilities, and through training.

Responsibilities and authorities are usually defined in the company's organizational charts and in job descriptions and/or specifications of departmental responsibilities. In addition, operational procedures usually define responsibilities for carrying out and supervising activities described in the procedures. Using these three types of documents it should be possible to completely define and document all responsibilities and authorities pertaining to the environmental management system.

DOCUMENTATION

The organizational charts may be included in the environmental system manual. In larger companies there is usually one chart depicting the general organization of the whole company and a separate chart identifying the functions having specific responsibilities in operating the environmental system.

Specifications of departmental responsibilities are also best documented in the environmental manual. They define for each department responsibilities related to the environmental management system. For example, relevant responsibilities of the Marketing department may be to research how customers feel about specific environmental aspects, to inform the market about the company's environmental management system and its achievements, and to research what is the general level of environmental performance in the industry and which companies are the leaders. Some departmental and/or functional responsibilities may also be specified in job descriptions.

Responsibilities and authorities related to specific activities are usually documented in procedures dealing with these activities. For example, a procedure dealing with management reviews should define who is responsible for scheduling and chairing the reviews, who should participate, and what is the authority of the review panel.

EVIDENCE/RECORDS

Evidence of conformance consists of documents defining the organizational structure and records demonstrating that the organizational responsibilities and authorities have been properly communicated within and outside the organization. Typically, these are document distribution records and/or training records.

AUDIT

Assessing conformance, auditors will review organizational charts, specifications of departmental responsibilities, job descriptions, operational procedures, and training records. It is important that someone on the auditee team can guide the auditors through these documents and point out where specific responsibilities and authorities are documented. For example, if the auditors ask who is responsible for conducting internal audits of the system and what is the authority of internal auditors, someone should be able to point out the specific charts, manuals or procedure where these responsibilities and authorities are defined.

Auditors will pay particular attention to the responsibilities and authorities of the top management. The documentation should clearly demonstrate that the top management is responsible for formulating the environmental policy, and for reviewing the system. In addition, to verify that the organizational structure is properly implemented, auditors may want to test whether the assigned responsibilities and authorities are indeed exercised. For example, if there is a requirement that decisions regarding disposal of hazardous waste must be authorized by VP Operations, auditors will check that in fact the VP is always involved.

While ISO 14001 does not explicitly require independence of personnel verifying environmental performance and auditing the environmental system, auditors will expect it, especially in larger organizations where there are dedicated environmental functions.

REQUIREMENT

Provide adequate resources for implementation and maintenance of the system.

The management is obliged to provide two kinds of resources: for developing, implementing, and maintaining the environmental system — mainly human resources and specialized skills; and for controlling and/or improving the environmental performance —

mainly technology and equipment resources. The need for resources for the environmental system depends on the size and complexity of the organization and the nature of its operations. The nature and magnitude of resources required to improve environmental performance is dictated by the goals set in the environmental policy and its objectives and targets. Although ISO 14001 does not specify any performance criteria, it requires that the organization commits itself — and allocates adequate resources — to reach its objectives and targets and thus fulfill its environmental policy.

DOCUMENTATION

To satisfy this requirement on documentation level it is sufficient to state in the environmental manual that the management is responsible for providing adequate resources for operating the environmental system and for improving environmental performance in line with the environmental policy.

EVIDENCE/RECORDS

No special records or other documentary evidence of conformance is required. There is no requirement for formal allocation of resources, such as budgets or dedicated staffing. Personnel qualification and training records may be relevant to demonstrate competence. The adequacy of allocated resources will be judged on the basis of the overall performance of the environmental system.

AUDIT

This is obviously an important requirement, but it is difficult to audit. Even when auditors can clearly see that there are insufficient staff and/or other resources to maintain the environmental system, they will usually find it difficult to point to specific objective evidence supporting their observation. In any event, to cite inadequate resources is very confrontational. Instead, auditors usually cite the nonconformances that are the result of insufficient resources, rather than directly citing the inadequacy of resources itself. For example, when corrective actions are not followed up in a timely manner due to inadequate staffing, the resulting nonconformance will usually just refer to the delay itself, without mentioning the underlying lack of resources.

REQUIREMENT

Appoint a management representative responsible for conformance with ISO 14001 and for reporting on the performance of the system.

The vice president, director, or manager responsible for environmental protection is a natural choice for the appointment as the management representative. But anyone in management can be appointed, irrespective of his or her other responsibilities. In companies where there are no functions and organizational units dedicated to environmental protection, it is often the health and safety or quality assurance manager that is appointed as the management representative. In small companies, it is often the president or the managing director.

The management representative does not necessarily need to have direct role in the day-to-day operation of the environmental system. His or her role is to assume the authority and responsibility for conformance and performance of the system. A common arrangement in smaller companies is to have the top executive be the management representative and a mid-level manager be responsible for the actual day-to-day operation of the system.

At a minimum, the management representative must be responsible for ensuring that the environmental management system is established, implemented and maintained to conform with requirements of ISO 14001; and for reporting on the performance of the system to the top management. When the environmental system is certified by a third-party registrar, the representative is also responsible for dealing with the registrar.

DOCUMENTATION

The appointment of the management representative must be formally documented. It is usually a statement in the environmental system manual. The appointment should state who is appointed (usually function rather than name) and define what is the authority and responsibility of the management representative. In addition, the function of the management representative should be depicted in the organizational charts. Some companies also like to document the position in job descriptions.

EVIDENCE/RECORDS

The evidence of conformance is the statement of appointment and records demonstrating that the management representative is indeed actively exercising his or her authority and assumes responsibility for the environmental system.

AUDIT

Auditors will review the statement of appointment of the management representative and the scope of assigned responsibility and authority. They will check whether the responsibility includes ensuring conformance with ISO 14001 and reporting to the top management; and whether the representative has sufficient authority to be effective. Auditors will also verify whether the management representative is indeed involved with the environmental system to the extent specified in his or her job description (or manual). The evidence will be provided by records of internal audits, corrective and preventive actions, management reviews, authorizations, and other documents resulting from, or recording activities of, the management representative.

4.4.2 Training, awareness and competence

REQUIREMENT

Identify training needs.

ISO 14001 requires that training be provided in the following three categories: 1) General awareness training about the environmental management system; 2) Training in emergency preparedness and response procedures; and 3) Specific technical and skill training for personnel whose work may create a significant environmental impact. The first two categories of training are usually administrated by the human resources department or other function responsible for company-wide training programs. The technical training is often provided internally by departments involved with handling, storage, or processing of hazardous substances, or other activities that may seriously impact the environment.

ISO 14001 requires that companies identify their training needs. Some basic training needs are suggested by the standard — notably the three categories of training listed above. There also will be training needs arising directly from specific programs for achieving the environmental policy and its objectives. Additional training needs may be identified in response to system or performance failures (nonconformances and incidences), or be determined from periodical evaluations of personnel qualifications and performance. Training program should be periodically reviewed (by management reviews, for example) and be modified as required to account for new products, activities, technology, or regulations.

DOCUMENTATION

Although the standard does not explicitly require a procedure for identifying training needs and administrating the training system, there should be something in the system documentation explaining what kinds of training programs are provided; who is responsible for planning and administrating training programs; and how suitability and effectiveness of training is reviewed. Whether required or not, a procedure is the most effective medium for documenting the training system.

EVIDENCE/RECORDS

Records and documents demonstrating that training needs are identified consist of training programs, personnel performance evaluations, review reports, minutes of management review meetings, and so forth. Corrective and preventive actions that utilize training to improve performance or conformance are also relevant evidence demonstrating identification of training needs.

AUDIT

Assessing conformance, auditors will investigate the processes for establishing and updating training programs. They will verify that there are functions responsible for identifying training needs and establishing training programs; that the programs are documented; that there are systems for evaluating personnel qualifications and performance; and that there are methods for reviewing suitability and effectiveness of training. Auditors will also review the historical evolution of training programs to assess whether they are indeed being updated in response to new developments and requirements.

REQUIREMENT

Make employees aware and knowledgeable of the environmental policy, significant impacts and the environmental management system.

Most of Section 4.4.2 is dedicated to requirements for employee awareness training. At a minimum, employees must be aware of:

- Importance of the environmental policy, procedures, and the environmental management system;
- Actual and potential environmental impacts of their work and benefits of improved personal performance;
- Their roles in achieving the environmental policy and maintaining the environmental system; and
- Potential consequences of deviating from procedures.

The standard requires that there be a procedure (i.e., a system) for developing awareness in the above-listed areas. A program for developing awareness may comprise training, distributing the environmental system documentation, distributing pamphlets, displaying posters and banners, posting messages on computer networks, playing video presentations, and so forth. There may also be programs for recognizing exceptional performance and achievement. While there are many possibilities, training is usually the main element. It may be classroom instruction, on-the-job demonstration, a self-study course, or an external seminar.

When implementing a new environmental management system, it is most effective to develop employee awareness through a specialized training program that focuses exclusively on the environmental issues. Examples include classroom training or distribution of a self-study course explaining the environmental system and preparing employees for the certification audit.

In mature environmental management systems the environmental awareness training can be combined with other training programs. For example, explanation of the environmental system and policy can be included in the general orientation training for newly hired employees, and training about specific environmental impacts and consequences of deviating from procedures can be combined with departmental skill training.

DOCUMENTATION

ISO 14001 explicitly requires a procedure for making employees aware of the environmental system and issues. The procedure should define the training and other programs for developing the awareness. It should outline the scope and objectives for the programs, and the responsibilities for developing and implementing them; and should specify how training programs should be documented and how their implementation should be recorded and reviewed.

EVIDENCE/RECORDS

Training records are explicitly required in Section 4.5.3, *Records*. The required records consist of personnel qualification and training records, and documentation specifying the content, format and duration of training. A set of materials presented and/or distributed during training may be retained as record of training content.

AUDIT

Assessing conformance, auditors will review training programs and training records, and will interview personnel to test their knowledge of the environmental management system. Auditors will ask employees what are the major goals and commitments in the environmental policy; what significant aspects are controlled by the environmental system; which environmental procedures are relevant to their work or function; how they would report dangerous or otherwise nonconforming conditions; and so forth. In addition, auditors will ask employees how they would react to emergency situations and, where special skill training is relevant, test their knowledge of specific procedures and techniques — for example, handling of hazardous materials. For further discussion of emergency response and special skill training, refer to the next two requirements.

REQUIREMENT

Train all personnel in emergency preparedness and response procedures.

Programs for training emergency procedures may include several elements: general orientation about where to find site plans, emergency procedures, first aid equipment, etc., and how to recognize various types of alarms and how to respond; more specific training in how to manage and control particular emergency situations; and emergency drills (where practicable). Everyone should receive general orientation training and participate in drills, and selected employees should be trained specifically in emergency functions assigned to them; for example, organizing evacuation, contacting authorities, operating emergency equipment, temporary containment, and so forth. When appropriate, environmental emergency training may be integrated with fire, earthquake, and other disaster response training and drills.

DOCUMENTATION

On the documentation level, there should be a procedure for training personnel in emergency response. The procedure should define the scope and nature of such training; instruct how to document the training content and how to establish and maintain training records; and assign the responsibility for these activities. The procedure is usually documented in a special section of the general training procedure.

EVIDENCE/RECORDS

The evidence of conformance consists of a training program or schedule, documentation of training content, and training records. Also relevant are records demonstrating qualifications of instructors.

AUDIT

Assessing conformance, auditors will review environmental preparedness and response procedures and the corresponding training and drill records to verify that all employees have been trained in the procedures. They may also interview some employees to test their knowledge of these procedures — especially employees that have specific roles and responsibilities in the emergency response programs.

REQUIREMENT

Train and/or ensure competence of personnel whose work may create a significant environmental impact.

This requirement relates to job-specific training of employees involved in environmentally sensitive operations and activities. For example, those responsible for receiving, storing, handling and disposing of chemicals and hazardous substances; maintaining environmentally related equipment and machinery; monitoring environmental indicators; operating processes that involve significant environmental aspects; and so forth.

Organizations must identify all operations and activities that may create a significant environmental impact, and define the level of competence required for specific functions and positions related to those operations. There must also be records of appropriate education, experience and training to demonstrate that the deployed personnel meets these competence requirements. Although not explicitly required in ISO 14001, job descriptions are very helpful in effective implementation of this requirement.

The amount of training required for a specific job will depend on the complexity of the job and the level of education and experience of the employee. In some instances there will be also laws and regulations prescribing specific training requirements.

DOCUMENTATION

An operational procedure should assign the responsibility and provide instructions for identifying all functions and positions that may cause significant environmental impacts; and for defining qualification requirements for these functions. There should also be instructions for establishing and implementing training programs to ensure that the qualification requirements are met by all deployed personnel. This procedure may either be established as a separate document or be included in the general training procedure.

EVIDENCE/RECORDS Records and evidence of conformance consist of a list or other documentation identifying functions and positions that can cause significant impacts; of job descriptions and/or qualification requirements for these functions; and of personnel qualification and training records. There should also be documentation specifying the content, format and duration of training. A set of materials presented and/or distributed during training may be retained as record of training content.

AUDIT Auditors will review the list of functions and positions that can cause significant impacts and will assess its completeness. They will also ask for job descriptions or other documents specifying required competence level for those functions, and will verify that there are appropriate training programs to ensure adequate competence of deployed personnel. To test whether the deployed personnel is indeed adequately trained, auditors will observe and interview employees performing environmentally sensitive functions and will review their qualification and training records. Auditors may also ask for a demonstration of a relevant activity, for example, dispensing a chemical or emptying a hazardous waste bin.

4.4.3 Communication

REQUIREMENT ***Establish systems for internal communication regarding environmental issues.***

To function properly, an environmental management system must include processes for communicating throughout the organization the policies, objectives and procedures of the system, and for reporting on the effectiveness of the environmental system and the environmental performance. The communication should be two-way. The management should communicate to the organization information and instructions about the significant environmental aspects; the environmental policy, objectives and targets; the environmental system procedures and consequences of deviating from these procedures; the methods and techniques to be used in operations and emergency situations; and the commitment of the top management to the environmental system. The organization should communicate back to the management (or those responsible for the environment) information and data concerning the environmental performance; the progress in achieving objectives and targets; the effectiveness of the environmental system; and specific concerns related to the environment. The communication may be in the form of written policies and procedures, memoranda, training, bulletin board postings, meetings, reports, records, and so forth.

This requirement does not in itself introduce any new obligations. It merely summarizes and reinforces other communication-related requirements stated elsewhere in the standard. Elements of the internal communication systems are already required in many other sections. For example, there are explicit requirements for communicating the environmental policy, for developing awareness about the environmental system, for documenting the system in procedures, for training, for maintaining environmental performance records, for reporting nonconformances, for recording management reviews, and so forth. A properly designed and implemented system will in itself comply with this requirement for internal communication without any additional actions.

ISO 14001 explicitly requires that there be procedures for internal communications. This does not mean that there must be a separate and dedicated procedure. Internal communication procedures are usually incorporated into other procedures dealing with specific activities that include communication. For example, a procedure dealing with the environmental policy would instruct how to communicate the policy, and a training program would define what information is to be communicated through training. Although ISO 14001 dedicates a unique subsection to internal communication, it is not a unique and easily identifiable element of the system. Communication activities are distributed throughout the whole system.

Documentation

Such distributed element is best described in the environmental manual. Being the top policy-level document, the manual can pull together and explain the general structure and aspects of the internal communication system. The manual should explain the purpose and importance of internal communication; encourage two-way communication (i.e., feedback regarding environmental performance and effectiveness of the system); provide guidance for considering communication when developing and documenting various elements and activities of the environmental system; outline other communication programs; and reference those procedures that further explain the internal communication systems, for example, document control, management review, record keeping, and training procedures.

Evidence/Records

Anything that demonstrates that information and data are communicated internally may be used as evidence of conformance with this requirement. These can be procedures, reports, memoranda, minutes of meetings, environmental performance records, training records, and so forth. Document distribution lists and records are also important to demonstrate that the relevant information is indeed communicated to appropriate functions.

Audit

Assessing conformance, auditors will first assure themselves that there are documented policies and guidelines for internal communication, and that the actual communication systems are defined in procedures. Being distributed throughout the whole environmental management system, the internal communication systems may be difficult to identify. External auditors who are not intimately familiar with the environmental system will need help. The company personnel escorting auditors should be able to point out specific sections in various procedures where the relevant information can be found. Escorts who are not prepared may not realize that there exists in their company an internal communication system and, when asked about such system, may not even understand what the auditors mean.

Implementation of the internal communication systems is audited continuously throughout the whole audit.

Whenever relevant to the audited activity, auditors will ask about the associated communication.

REQUIREMENT

Establish system for receiving communication from external interested parties.

This requirement applies to communication from external parties pertaining to the environment or the environmental management system. The external parties may be governmental agencies, authorities and public organizations concerned with emergency preparedness and response, registrars of ISO 14001 environmental management systems, customers, consumers, labor unions, organizations and individuals concerned with protection of the environment, community groups, neighbors, and so forth. The requirement calls for a system for receiving, documenting and responding to such communication.

The function responsible for handling external communication should be clearly defined. In larger companies it is usually the function of corporate communications and/or the function of environmental protection. In smaller companies where such dedicated functions do not exist, the management representative for the environment (refer to Section 4.4.1.) is usually responsible for external communication. The function charged with handling of external communication must be clearly depicted in the organizational charts and/or defined in specifications of departmental responsibilities.

All received communication should be logged and be forwarded to those responsible for the response. If the communication is verbal, it should be documented in a memorandum of the conversation or minutes of the meeting, as appropriate. There should also be a system for tracking progress and status of the communication and response process. It may be a log, or a computerized database, recording the nature of the communication, the function responsible for the response, and the status of the response (open or closed).

There should be a link between the system for processing external communication and the system for preventive and corrective actions. When an external party identifies an unacceptable condition that may need to be corrected or improved, the system should ensure that pertinent information is communicated internally to those responsible for considering and implementing corrective actions. There should also be a link with the system for establishing and reviewing environmental objectives and targets. Section 4.3.3 requires that views of interested parties be considered in setting the environmental objectives.

The system for receiving and processing external communication is inherently similar to the system for processing customer complains, required in ISO 9000. Those companies that already have a well-developed customer complaint processing system may just extend its area of application to handle the external environmental communication.

DOCUMENTATION

The standard explicitly requires an operational procedure for handling and processing external communication. The procedure should define the functions responsible for receiving and responding to communication from external parties, instruct how communication should be documented and logged, and explain the system for tracking and recording the response.

EVIDENCE/RECORDS

The records and evidence of conformance consists of receiving and tracking logs, and the letters, memos, minutes and reports that document the external communication. There should also be evidence of internal communication demonstrating that the issues raised by external parties are brought to the attention of relevant functions and that preventive and/or corrective actions have been initiated when appropriate.

AUDIT

Auditors will first verify that there is a documented and implemented system for receiving, documenting and processing external communications. They will assess whether the system ensures that communication will not be lost, that it will be forwarded to the appropriate responsible function, that a response will be processed without unnecessary delays, and that other functions in the organization will be informed about the communication when relevant. Auditors will usually choose a sample of cases and will review the files in detail. They will pay attention to dates and the level of authority making the response decisions.

Auditors will not judge whether a given response was appropriate or not unless the response, or lack of a response, obviously violates policies, procedures or legal obligations.

REQUIREMENT

Consider external communication of significant environmental aspects.

This requirement obliges organizations implementing ISO 14001 to consider whether they want to communicate to the public the information their significant environmental aspects. The decision must be formally recorded.

This requirement is a compromise between the European approach (EMAS), which requires that the environmental aspects and performance of an organization be available to the public, and the concern of the U.S. delegation that such requirement may be in conflict with legal considerations in the United States. As a compromise, ISO 14001 mandates only publication of the environmental policy, while the publication of significant environmental aspects is voluntary. There is no mention of any obligation to publish the environmental performance data.

DOCUMENTATION

The decision whether significant environmental aspects will be communicated to the public must be recorded. The best place for documenting the decision is the environmental system manual. If the decision is to keep the aspects confidential, the manual should state that the environmental aspects are confidential and may not be communicated outside the organization without prior approval of the president (or other designated function). If the decision is to make the aspects available to the public, the manual should instruct how they should be communicated. The environmental aspects are usually communicated in the same way, and together with the environmental policy.

EVIDENCE/RECORDS If the decision is to communicate the significant aspects, the evidence of conformance consists of the document, internet site, press release, or other medium communicating the aspects. Otherwise, the evidence consists of a documented statement that the significant aspects will not be communicated to the public.

AUDIT Auditors will look for a formal declaration stating whether the significant environmental aspects will or will not be communicated to the public. Irrespective of the decision, conformance with this requirement is achieved as long as such statement is included in the environmental system manual or is otherwise documented. If the decision is to publish the aspects, auditors will verify that they are indeed communicated in accordance with the applicable procedure.

4.4.4 Environmental management system documentation

REQUIREMENT ***Describe the environmental management system in a manual and reference the related documentation.***

ISO 14001 requires that the core elements of the environmental management system and their interrelation be documented. At a minimum, the documentation should consider the four elements corresponding to the sectional headings in the standard — *Planning, Implementation and operation, Checking and corrective actions,* and *Management review* — and provide direction to the related documentation, such as the environmental policy; the significant environmental aspects and impacts; the environmental objectives and targets; the management programs for achieving environmental objectives; assignments of key roles and responsibilities; operational procedures; emergency response plans; process information; and job instructions.

DOCUMENTATION

The general description of the system is usually documented in the environmental system manual. The manual is the top-level document in a three-level documentation system. The second-level documents are the operational procedures. The third-level documents are job instructions, specifications and standards. This is a classic documentation model used for quality management systems, including ISO 9000. The role of the environmental manual is to state the environmental policy; to outline the structure and to summarize the major procedures of the system; to define the principal responsibilities and authorities; and to reference other documents supporting the system, especially the operational procedures.

Much of the information contained in the environmental manual will be repeated and further developed in the operational procedures. For this reason, many organizations do not see any added value in having the manual as a separate document, and either combine it with the operational procedures or reduce its contents to meaningless paraphrasing of ISO 14001 requirements. But even though the manual may look just like a summary of the operational procedures, it has an important independent role. Within the organization, it defines the core principles and policies for developing and maintaining the environmental system and its documentation. For external parties, it is a short and concise document providing general information about an organization's environmental management system.

A well-written and presented environmental system manual is a powerful marketing and public relations tool. It can be enclosed with offers and be otherwise distributed to customers, governmental agencies, community representatives, and other interested parties. To serve this purpose well, the manual should be brief but sufficiently developed to eloquently present the environmental system. It should also project the environmental culture and image of the company.

The environmental manual is usually divided into the same four main sections as the standard. Subsectional divisions may be also the same, or similar. The manual usually starts with document identification and control pages, a short introduction, the environmental policy, and organizational charts and specifications of responsibilities and authorities. A list of operational procedures may be included in the back of the manual. The elements of the environmental system should be described broadly, focusing on policies and assignment of key responsibilities rather than detailed instructions. The manual should also introduce and reference the operational procedures, plans, instructions, management programs and other associated documentation.

EVIDENCE/RECORDS The evidence of conformance is the environmental manual itself and a record of its approval and release.

AUDIT Auditors usually review the environmental manual, or other form of top-level documentation, during the initial on-site visit (Phase I audit). The review is one of the criteria used to judge whether an organization is ready for the Phase II implementation audit. To be acceptable, the manual (or other documents) must outline all elements of the system, define main responsibilities, and provide direction to the related documentation.

4.4.5 Document control

REQUIREMENT ***Define the document control system and the types of controlled documents.***

The scope of application and structure of the document control system should be well explained, so that it is clear what type of documents must be controlled and who is responsible for controlling them. This is especially important in organizations where the document control system is not centralized and where each department issues and controls its own documents.

The first sentence in ISO 14001 Section 4.4.5 defines the scope of application for the document control system to include "... all documents required by this standard ...". Auditors interpret this definition to comprise all documents established to define, implement, and maintain (operate) the environmental management system, whether they are explicitly named in the standard or not. Examples include environmental system manual; environmental policy; significant environmental aspects and impacts; environmental objectives and targets; management programs for achieving environmental objectives; assignments of key roles and responsibilities; operational procedures and job instructions; process information; training programs; environmental performance specifications; applicable legislation and regulations; emergency response plans; and so forth. In short, any document that contains instructions, specifications or data pertaining to products, processes or activities that are affected by the environmental management system should be controlled.

The control requirements apply equally to documents and data stored in electronic media, such as computer files, databases, and so forth.

DOCUMENTATION

General presentation of the document control system may be included in the environmental manual or, in larger organizations, in a separate operational procedure. The categories of controlled documents are usually defined in the manual. Individual procedures and instructions explaining how to control specific types of documents should also define the categories of documents to which they apply.

EVIDENCE/RECORDS

The evidence of conformance is a list specifying the types of controlled documents. The list can be included in the environmental manual or a procedure.

AUDIT

Auditors will verify that all relevant documents are included in the document control system. There is never a problem with the environmental manual and operational procedures. However, companies often forget to include in the document control system such documents as handling, storage or labeling instructions, training programs, emergency response plans, or external regulations and standards.

REQUIREMENT

Approve documents for use and periodically review them, and revise as necessary.

Every document must bear a signature(s), or other identification, of the authorized personnel who reviewed and approved the document for use. If more than one signature is on a document, it should be clear which signatures evidence review and which evidence approval. The approval signature usually distinguishes an issued document from its draft or preliminary versions. For each type of document, the authority to approve and issue documents must be defined. Document revisions, corrections and reissues must be likewise reviewed and approved.

Documents must be reviewed periodically. In practice this means that documents may not be valid indefinitely without periodical review confirming that they are still relevant and adequate. For example, validity of environmental procedures may be limited to three years — i.e., this would be the maximum period of time allowed between two consecutive revisions — and then procedures be recalled for a review and reissued on a higher revision level, whether they have been changed or not.

DOCUMENTATION

The general document control procedure, or specific procedures for various types of documents, should have a section dealing specifically with review and approval of documents. The procedure(s) should explain the manner and scope of review, and define the level of authority required to approve and issue various types of documents. There should be also instructions specifying how, and how often, various types of documents must be reviewed for continuing adequacy.

EVIDENCE/RECORDS

Signatures on documents provide the required record and sufficient evidence of review and approval. For documents in electronic media, the signatures may be substituted by initials or other identification of those who reviewed and approved the documents. The evidence that documents are being periodically reviewed is provided by document review and approval dates.

AUDIT

Conformance with this requirement is verified not only in the locations where documents are issued and controlled, but also everywhere else they are used and stored. When examining documents, auditors always look for review and approval status. It is a reflex. Auditors will also note the initial issue and revision dates to verify that documents are regularly reviewed for continuing adequacy.

REQUIREMENT

Specify retention periods for controlled documents.

ISO 14001 explicitly requires that controlled documents be retained for a specified period of time. It is still not entirely clear what this requirement means. Active documents must, of course, be retained while they are active, and it would not make much sense to specify retention periods for active documents. Following this reasoning, the requirement must apply to historical documents and obsolete documents that are superceded by new revisions. The requirement to retain such documents makes sense in cases where it is important to preserve the knowledge how things were done in the past.

Not all historical documents must necessarily be retained, and the retention periods may be as short or as long as dictated by actual need and legal and other considerations. ISO 14001 does not prescribe any specific periods, and auditors will normally not question the specified periods, unless they do not make any sense.

DOCUMENTATION

The document control procedure should have a section dealing with document retention periods. It should instruct that retention periods for all controlled documents must be established, and it should either provide the specific retention periods for various types of documents or instruct how retention periods should be otherwise specified.

EVIDENCE/RECORDS

The evidence of conformance are the document retention periods specified in procedures, or in the documents themselves.

AUDIT

Assessing conformance, auditors will verify that retention periods are specified for all types of controlled documents. Auditors will usually accept the specified retention periods unless there is an obvious problem with a document being discarded too quickly, or when there is a conflict with legal or other requirements. To test implementation, auditors will ask for retrieval of some older documents from various categories.

REQUIREMENT

Identify documents with their revision level, and maintain a master list with current revision status of documents.

Documents should be directly identified with their revision level. Consecutive numbering using numerals or letters is the most common method, but in some cases it may be appropriate to include the revision level in the code identifying a document, or simply use the date of issue as indication of the revision level.

There must also be a system for verifying what is the current revision level of a given document. The most common method is a master list of issued documents. To satisfy this requirement, the list need only specify the latest revision of a document; however, it could also contain such additional information as the date of issue, identification of the issuing authority, distribution of the document, and its retention period. This would help to satisfy other document control requirements. Document master lists can be maintained on computer databases, catalog cards, or manual logs.

DOCUMENTATION

The general document control procedure should have a section dealing with identification and verification of document revision status. The procedure should instruct how revision levels are identified for various types of documents, and prescribe specific means for tracking their revision status — a master list, for example.

EVIDENCE/RECORDS

The records and evidence of conformance are provided by the documents themselves, displaying their revision level (or date of issue); and by master lists providing the means for verifying the current revision status of documents.

AUDIT

Assessing conformance, auditors will check if documents display their revision levels, and will review the master lists or logs of issued documents. Auditors will also test the system for verifying revision status. They will pick a sample of documents from different locations around the company, note their revision levels, and ask for the verification that these revision levels are indeed current.

REQUIREMENT

Ensure that appropriate documents are available at all locations where they are needed.

There are two issues in this requirement. One is identification of the locations where specific documents should be available, and the other is effective distribution of the documents. For documents having permanent or fairly constant distribution, there should be documented distribution lists. The lists can be printed on the documents themselves or be maintained separately.

When distribution cannot be predicted in advance, documents can be made available on an as-needed basis. But their use must be recorded so that it is known where to follow up with revisions. If maintaining such a record is not practical, documents should be destroyed promptly or returned after one-time use. Any filing or storage of documents that cannot be automatically replaced with new revisions must be prohibited.

DOCUMENTATION

The general document control procedure, or specific procedures dealing with various types of documents, should have a section dedicated to distribution of documents. The procedure should explain how distribution lists are documented and maintained, and what means are used to physically deliver documents to their destinations. It should also explain, if applicable, how to use letters of transmittal, sign-off logs, or other means for confirmation of receipt. This procedure can be combined with the procedure regulating withdrawal of obsolete documents from points of use (see next requirement).

EVIDENCE/RECORDS Distribution lists are the main evidence that the locations where documents should be placed are identified. The actual distribution of documents can be recorded in letters of transmittal, distribution logs, or simply by checking off names on a distribution list. Any method is acceptable as long as it is documented in procedures. Establishment and maintenance of receipt records is not explicitly required.

AUDIT Assessing compliance, auditors will review the distribution lists and records, and test the distribution system. This can be done by either picking a name from a distribution list and verifying that the person indeed has the document; or the other way around, by picking up a document from somebody's file and verifying that the person is on the document's distribution list.

REQUIREMENT ***Remove obsolete documents from points of use and identify retained historical copies of obsolete documents to preclude unintended use.***

Obsolete documents are usually removed from points of use at the same time that new revisions of the documents are distributed. There are basically two possibilities: either superseded documents are exchanged for revised versions, or the recipients of revised documents are obliged to destroy the old copies. Direct exchange is more effective, of course, but the other system is also acceptable, provided that it can be satisfactorily implemented.

When historical documents need to be retained for preservation of knowledge and/or legal reasons, they should be stamped OBSOLETE or HISTORY, and be kept separate from current documents to preclude unintended use.

DOCUMENTATION The general document control procedure, or specific procedures dealing with various types of documents, should have a section dedicated to withdrawal and handling of obsolete documents. The procedure should explain the method used for withdrawal of obsolete documents and prohibit unauthorized filing and storage of any obsolete documents and any documents that are at risk of becoming obsolete because they are not automatically updated with new revisions. The procedure should also explain how historical copies of obsolete documents are identified and stored.

EVIDENCE/RECORDS

There are normally no records or other written evidence that obsolete documents are being withdrawn from points of use, unless a document exchange system involves logging the withdrawn documents. Handling and storage of historical copies of obsolete documents is evidenced by proper identification of the documents themselves.

AUDIT

Because there are generally no records or other written evidence, the conformance will be assessed by testing the system. Auditors will check documents against their master lists and will look out for multiple revisions of a document in files and points of use. They will also review the files of obsolete historical documents.

REQUIREMENT

Review and approve changes in documents and reissues of revised documents.

Review and approval of revisions involving reissue of a document on a higher revision level should follow the same procedure that applies to issuing initial documents (refer to the second requirement in this section).

Corrections in documents, often called red-line corrections, must also be reviewed and approved. Anything crossed out, added, or changed in a controlled document must be dated and signed by an authorized person. Red-line corrections should not be left on documents indefinitely. Documents should be reissued after a practical number of changes have been made.

Handwritten notes and comments on documents should be avoided. Even though they do not modify the original content of a document, they add new information that is not authorized and will be lost when the document is reissued and the marked-up copy destroyed.

DOCUMENTATION

The procedure dealing with document changes should define the channels for requesting changes in documents; provide rules for making corrections in documents, including their review and approval; and instruct how documents should be reissued on a higher revision level.

EVIDENCE/RECORDS

The records and evidence of conformance are the signatures authorizing corrections and reissues of revised documents.

AUDIT

Any handwritten corrections and notes on documents will alert auditors and provoke questioning. Signed-off corrections are acceptable, but the sign-offs must be authorized and all concerned functions must be notified about the correction.

4.4.6 Operational control

REQUIREMENT

Identify and specify those operations and activities that are associated with significant environmental aspects.

Operations and activities associated with significant environmental aspects are usually identified at the stage of considering the environmental aspects and rating their significance. Relevant activities are also identified and reviewed when establishing environmental objectives and targets, and many of them are planned and documented in conjunction with development of management programs for achieving the objectives. Thus, most of the documentation and implementation work associated with this requirement is carried out at the stage of planning the environmental management system. The environmental aspects, objectives, targets and the associated management programs are discussed in this book under Section 4.3, *Planning*.

The types of operations and activities that should be considered are design and engineering, production processes, maintenance, handling, storage, transportation, purchasing, marketing, customer service, acquisition or construction of facilities, and so forth. The identified operations and activities should be listed or be otherwise documented. The list may be in the form of a matrix associating significant environmental aspects with the corresponding activities.

The next step is to ensure that the identified activities are carried out under specified conditions. Personnel involved with the activities must have the relevant process information and data, and must be qualified and instructed or trained in carrying out the activities. There must be a clear assignment of authority and responsibility. In many cases written procedures and instructions will be necessary (refer to the next requirement).

DOCUMENTATION

There is usually no need for a special procedure explaining how to identify the operations and activities associated with significant aspects and how to ensure that they are carried out under specified conditions. It is sufficient to state these policy-level principles in the environmental system manual. When required, more detailed instructions may be included in procedures. For example, procedure for selecting and rating environmental aspects may include instructions to also identify and document operations and activities associated with the aspects.

EVIDENCE/RECORDS

Evidence of conformance consists of a list or a matrix identifying relevant operations and activities, and documents specifying criteria and conditions for these activities. There should also be appropriate personnel qualification and training records and, when relevant, equipment maintenance records.

AUDIT

Verifying conformance, auditors will compare the list of significant environmental aspects with the list of associated operations and activities. Normally there should be at least one activity for each aspect — per definition, an aspect arises out of an activity. Then, auditors will select a number of activities and verify that they are carried out under specified conditions. Auditors will review work instructions and other documentation pertaining to the activity, personnel qualification and training records, and records demonstrating environmental performance resulting from the activity. Auditors will also interview personnel involved with the activity and, when possible, witness the activity itself.

REQUIREMENT

Establish operational criteria, procedures and instructions to ensure conformance with the environmental policy, objectives and targets.

There is no consensus on the scope of application of this requirement. Some auditors apply it narrowly to only those operations and activities that correspond to significant environmental aspects, while others apply it broadly to all activities related to the environmental management system, including activities for defining and administrating the system itself; for example, establishing environmental objectives, document control, or internal auditing. The difference is fundamental. The narrow interpretation limits the application to such third-level documents as work instructions and process procedures, while the broad interpretation creates a mandate to document all elements and activities of the environmental management system. As the standard is generally vague in requiring documented procedures, the broad interpretation is a powerful tool in forcing organizations to document their systems.

The narrow interpretation seems to be the correct one. Although the requirement in point a), that: "documented procedures be established and maintained to cover situations where their absence could lead to deviations from the environmental policy and the objectives and targets", sounds like a very broad requirement for documenting the whole environmental system, the first sentence in Section 4.4.6, *Operational control,* clearly indicates that the section applies only to activities associated with significant aspects.

Another problem with interpretation of this requirement is the uncertainty where exactly documented procedures are required. The standard does not suggest any criteria for determining whether absence of a procedure could lead to deviation form the environmental policy. Without a doubt, procedures are required where there is a history of actual problems. But there must be also more proactive criteria. Such factors as complexity of the operation, the level of operator training, and the potential consequences of failure should be considered when deciding whether a procedure should be written for a given operation.

DOCUMENTATION There is no need for a special operational procedure for writing work instructions. It is sufficient to make a policy-level statement in the environmental system manual that process criteria and work instructions shall be established where their absence could lead to deviations. The manual should also assign the responsibility for establishing and maintaining the instructions, and may outline the criteria to be used in determining when written instructions are needed.

EVIDENCE/RECORDS Evidence of conformance consists of procedures, work instructions, data sheets, operating and maintenance manuals, and other such documents instructing how to carry out critical operations and activities. Personnel qualification and training records are also relevant, especially where there are no written instructions.

AUDIT Auditors will expect to see some written procedures for most operations associated with significant aspects. Undocumented operations will invite questioning. Where there are no written procedures, auditors will be asking how personnel know what to do in specific situations and may want to cross-examine several employees to verify that everyone has been instructed to do the same thing. Another approach used by auditors when faced with an undocumented activity is to investigate its performance history. If there is a history of problems traceable to operator errors, the lack of written procedures is an automatic nonconformance. Undocumented critical operations and activities are difficult to defend, unless they are simple and obvious.

REQUIREMENT

Communicate environmental procedures and requirements to suppliers of products and services having significant environmental aspects.

Through this requirement, ISO 14001 attempts to include purchasing activities into the scope of the environmental management system. It is a weak and ambiguous requirement that may be difficult to audit and enforce.

In Section 4.3.1, *Environmental aspects,* the standard requires that, in addition to their own direct aspects, organizations identify and evaluate those environmental aspects over which they can be expected to have an influence. One category of such aspects are the environmental aspects of purchased products and services. The weakness of the requirement is that, while every organization should be expected to identify some aspects in this category, there will be many legitimate cases where none of them will be classified as significant. In such cases this requirement is technically unenforceable.

There is no intent in this requirement to extend the environmental management system to suppliers. An organization implementing ISO 14001 does not have to request its suppliers to improve their environmental performance, or even comply with applicable environmental laws. Companies should encourage suppliers to implement their own environmental management systems, but there is no mandate that they do. The intent of this requirement is limited to controlling purchased products and services rather than influencing environmental aspects of suppliers operations. Ultimately, the system must ensure that the products and services used by an organization do not compromise its environmental policy.

The environmental procedures and requirements contemplated in this section may be, for example, requirements for warrants certifying absence or limited concentrations of certain substances in products, requirements for special product packaging and/or labeling, procedures for delivering and unloading special categories of products, procedures for subcontracted cleaning and maintenance, and so forth. Contractors performing services on the premises may also be required to demonstrate adequate personnel qualification and training, satisfy licensing requirements, familiarize themselves with emergency procedures, and otherwise conform with relevant requirements of the environmental management system.

DOCUMENTATION

Although not explicitly required, there should also be an operational procedure instructing purchasing, and other functions responsible for specifying and purchasing products and services, how to qualify products and suppliers so that product characteristics and supplier operations are compatible with the organization's environmental policy. The procedure should define a system for gathering and evaluating information about supplier's environmental management, performance and compliance, and it may outline some general criteria for qualifying and/or rating suppliers. There should also be a documented system for defining, reviewing and verifying requirements with regard to environmental aspects of purchased products and services, including packaging, labeling, handling and disposal instructions, and so forth. Companies operating quality management systems have similar procedures for evaluating quality capabilities of suppliers.

EVIDENCE/RECORDS

Records and evidence of conformance consist of documents communicating to suppliers the environmental requirements. These can be product specifications, purchase orders, work instructions, etc. In addition, there should be some evidence that the organization is concerned with general environmental performance of its suppliers. These can be questionnaires, lists of approved suppliers, and communication encouraging suppliers to implement their own environmental management systems.

AUDIT

Evaluating conformance, auditors will verify that relevant environmental requirements are specified for purchased products and services, and are communicated to suppliers. Typical areas of concern are chemical substances and their packaging and labeling; services related to protection of the environment (for example treatment or disposal of waste, or recycling); and services performed on the premises (for example, cleaning, maintenance or construction).

Auditors will also be interested in a broader system for purchasing control. They will ask about supplier evaluation and rating, and will want to hear what is being done to encourage suppliers to improve their environmental management and performance.

Finally, auditors will verify that there is at least a good faith effort made to extend the control of significant aspects to suppliers. For example, if an organization has an objective to reduce the use of packaging material, auditors would expect that suppliers are also encouraged to review their own packaging specifications as well.

4.4.7 Emergency preparedness and response

REQUIREMENT

Identify potential for accidents and emergency situations.

Responding to this requirement, organizations should start with inventorying all hazardous substances they use, store or transport; including identification of quantities and locations. The next step is to identify potential situations when these substances could be accidentally released and, when relevant, to model the dispersion of hazardous gases and liquids. The last step is evaluation of potential impacts of an accident on surrounding areas.

Where certain types and quantities of hazardous substances are stored or handled emergency planning activities are required by law — EPCRA, for example. If a facility is subjected to, and complies with, legal and regulatory requirements with regard to emergency planning, it probably also satisfies all ISO 14001 requirements under this section. However, conformance with this ISO 14001 requirement must be demonstrated irrespective of legal and regulatory obligations. Facilities that are below the quantity threshold, or for any other reason are not subjected to, or are exempted from, the law must still demonstrate that they have have a procedure for identifying potential accidents and emergency situations.

DOCUMENTATION

ISO 14001 explicitly requires a procedure for identifying potential accidents and emergency situations. The procedure should assign the responsibility and provide instructions for inventorying hazardous substances; identifying potential accidents; modeling potential dispersion patterns; identifying environmentally sensitive features and facilities in the neighborhood (residential communities, schools, hospitals, watercourses, wildlife habitats, etc.); and evaluating potential impacts. The procedure should also require a periodical review of identified potential accidents and emergency situations, to account for new hazardous substances and other changing circumstances.

EVIDENCE/RECORDS

The record and evidence of conformance with this requirement usually consists of chemical inventory lists, material safety data sheets, site plans, neighborhood maps, scenarios of potential accidents, and reports evaluating potential impacts.

AUDIT

Auditors will want to see evidence that potential accidents and other emergency situations have been identified and evaluated, irrespective of legal and regulatory obligations. Where laws and regulations pertaining to emergency preparedness planning do not apply, auditors will still insist that there be a process for identifying and evaluating potential emergencies. In some organizations the risks for any potential emergency will be extremely small and there will be no need to develop formal emergency plans, other than designating evacuation routes. When merited, auditors will accept such conclusion; however, it must be based in a systematic evaluation and there must be evidence that such evaluation was carried out.

REQUIREMENT

Establish emergency preparedness and response procedures, and test the procedures where practicable.

Having identified and evaluated potential accidents and emergency situations, organizations must develop appropriate procedures to deal with these emergencies. The scope and extent of emergency procedures will depend on the nature of identified risks and, possibly, legal and regulatory requirements. Emergency preparedness and response procedures may include site plans; evacuations plans; assignment of responsibilities for managing emergency situations; shut-down instructions; manuals for use of personal protective equipment, decontamination equipment, communication devices, etc.; instructions for calling outside assistance; phone numbers, names and organizations that need to be notified; and so forth. The procedures must be generally accessible and personnel must be trained in their use (refer to Section 4.4.2, *Training*).

Emergency preparedness and response procedures should be periodically reviewed and, when necessary, revised to account for new hazards and changing circumstances, such as substantial change in the chemical inventory, introduction of new processes, and changes in the neighborhood. Adequacy of the procedures should be also evaluated following the occurrence of accidents and emergency situations, and experiences from test drills.

Emergency procedures should be periodically tested where practicable. In this context, practicable means possible rather than convenient. When it is not practicable to test or drill the whole emergency response system, those elements that can be tested should be identified and be regularly tested. For example, even when full scale evacuation drills are not possible, protective equipment, communication systems, and other equipment and systems should be tested at prescribed frequencies. When appropriate, testing of environmental emergency procedures may be integrated with testing of fire, earthquake, and other disaster response procedures.

DOCUMENTATION

On the policy level, the environmental system manual should include a commitment to establish emergency procedures; to implement them through distribution and training; and to maintain the procedures through testing and reviews. The actual emergency preparedness and response procedures must be documented and be accessible to all personnel. They may be in a form of written procedures, site plans, signs, posters, and so forth. The procedures do not necessarily have to be issued and controlled by the environmental management system. They may be established by the same department or authority that issues fire, earthquake or other emergency procedures. In fact, the environmental emergency procedures would normally be merged with other emergency procedures.

EVIDENCE/RECORDS

The evidence of conformance consists of the emergency procedures themselves, and of records demonstrating that the procedures are regularly tested and reviewed.

AUDIT

Assessing conformance, auditors will verify that emergency procedures exist and cover for all identified potential accidents and emergency situations. They will also verify that emergency procedures are available to all personnel and that everyone knows where they are placed; that, when practicable, the procedures are periodically tested and/or drilled; and that they are reviewed and, when necessary, revised to account for changing circumstances. A typical nonconformance, for example, would be a failure to review emergency procedures in connection with construction of a new warehouse.

4.5 CHECKING AND CORRECTIVE ACTION

ISO 14001 Section 4.5 defines those elements of the environmental management system for verifying and recording the environmental performance and the effectiveness of the environmental management system. There are also elements for identifying problems and implementing corrective actions.

Specifically, this section deals with systems for measuring and monitoring processes, activities and operations; for evaluating regulatory compliance; for identifying and dealing with nonconformances; for implementing corrective and preventive actions; for establishing and maintaining environmental records; and for conducting internal audits of the environmental management system.

The purpose of these elements is to verify whether the operational systems, programs and controls that were implemented to achieve the environmental policy are effective. This is the third link in the Plan-Implement-Check-Review cycle.

All elements in this section have their equivalents in ISO 9000. Requirements with regard to corrective and preventive actions, records, and internal auditing are almost identical. The first subsection, *Monitoring and measurement*, is equivalent to ISO 9000 requirements for inspections and management of inspection and test equipment. The only requirement that has no direct equivalent in ISO 9000 is self-evaluation of regulatory compliance (although it is also a form of inspection or audit). Those companies that already operate an ISO 9000 quality management system may draw on their experience with these elements, and even combine some of the procedures.

4.5 Checking and corrective action

4.5.1 Monitoring and measurement

REQUIREMENT

Monitor and measure key characteristics of those operations and activities that are related to significant aspects.

The purpose of this requirement is to ensure that there is a system for tracking environmental performance in relevant operations and activities. In particular, performance must be monitored for those characteristics that have been identified as significant aspects.

For those significant environmental aspects that are translated into objectives, and where the objectives are further developed into specific, quantifiable targets, it is sufficient to establish monitoring and measuring programs for the targets, and the significant aspects and objectives will be covered automatically. Monitoring and measuring is most appropriate on the target level because targets are specific and quantifiable. However, when a significant environmental aspect is not followed up with a specific objective and target, the monitoring and measuring programs must be developed directly for the aspect. A significant aspect may not be left out without any monitoring, whether it is followed up with an objective or not.

A program for monitoring and/or measuring of a characteristic should be defined in terms of the parameter to be measured (quantity, concentration, number of incidences, etc.); measuring or monitoring method; measuring equipment to be used; frequency and sampling method; reporting format; and assignment of the function responsible for taking the measurements and reporting results.

DOCUMENTATION

The standard explicitly requires that there be documented procedures defining the monitoring and measurement programs. The documentation can be on two levels: an operational procedure defining the general system and responsibilities, and specific plans and work instructions for monitoring and measuring particular characteristics.

The operational procedure should define the criteria for choosing the characteristics to be measured; describe general requirements for documenting the methodology, and for recording results; and assign responsibilities for carrying out various types of measurements and monitoring programs. The plans and work instructions should include a list or matrix of all characteristics that are being monitored; and detailed instructions with regard to frequency, sampling, use of measuring equipment, and reporting requirements for each characteristic. When appropriate, there should be documented acceptance limits, and instructions on what to do when limits are exceeded.

EVIDENCE/RECORDS

The records and evidence of conformance consists of procedures, plans and instructions, and logs or other records with the actual measurement or monitoring data. There should also be evidence that the data is regularly reviewed for trends and that appropriate response actions are taken to correct or improve performance.

AUDIT

Auditors will first compare the list of significant aspects, objectives and targets against the existing monitoring and measurement programs, to verify that all relevant characteristics of operations and activities are included. Then they will review specific programs, verifying that measurements are taken at the prescribed frequency, that there is a clear assignment of responsibility for taking the measurements, and that there are sufficient instructions. Auditors will also review the monitoring and measurement data and will verify that, when appropriate, the data is indeed checked for compliance with established criteria and analyzed for trends.

REQUIREMENT

Periodically evaluate compliance with applicable environmental laws and regulations.

Self-evaluation of compliance with laws and regulations may have a very significant legal dimension. The legal aspects of this requirement are discussed in *ISO 14001 and the Law* by S. Wayne Rosenbaum. In this book self-evaluation of compliance is approached in strictly technical terms, focusing on conformance with ISO 14001 as understood by a systems auditor, rather than an attorney.

This requirement is very similar to the first requirement in this section, except that instead of monitoring performance with regard to significant environmental aspects, it obliges organizations to do the same for aspects related to legal and regulatory requirements. In cases when activities subjected to regulatory requirements are also classified as significant aspects, the same monitoring and measurement system may be used to satisfy both requirements.

The system for evaluating compliance should be defined in the same terms as the system for monitoring significant aspects. It should define the characteristic or regulation to be evaluated; parameters to be measured (quantity, concentration, number of incidences, etc.); measuring or monitoring method; measuring equipment to be used; frequency and sampling method; reporting format; and assignment of the function responsible for taking the measurements and reporting results.

The system for evaluating compliance must be applied to all relevant legislative and regulatory requirements. A list of applicable regulations should have been already compiled in response to an earlier requirement in Section 4.3.2 of the standard, *Legal and other requirements*.

DOCUMENTATION

The standard explicitly requires that there be documented procedures defining the compliance evaluation system. The documentation can be on two levels: an operational procedure defining the general system and responsibilities, and specific plans and work instructions for evaluating compliance with specific regulations.

The operational procedure will be very similar to the procedure that instructs in monitoring of environmental performance with regard to significant aspects (refer to preceding requirement). The only substantial difference is the scope of application of the procedure — to comprise monitoring of compliance with regulatory requirements instead of significant aspects. In most circumstances both procedures can be merged, unless organizational or legal considerations favor separate systems and procedures.

The plans and work instructions should include a list or matrix of all regulatory requirements against which compliance is being evaluated; and detailed instructions with regard to frequency, sampling, use of measuring equipment, and reporting requirements. The list of regulatory requirements should have been provided by the system for identifying legal requirements (see Section 4.3.2). Personnel evaluating compliance must know what the acceptable limits for the monitored parameter are, and what to do when the limits are exceeded.

EVIDENCE/RECORDS

The records and evidence of conformance consist of procedures, plans and instructions, and logs or other records with the actual measurement or monitoring data. There should also be evidence that the data is immediately checked for compliance and regularly reviewed for trends, and that appropriate response actions are taken to correct or improve noncomplying performance.

AUDIT

Auditors will first compare the list of applicable legislation and regulations against the established compliance evaluation programs, to verify that all applicable regulations are included. Then, they will review specific programs, verifying whether measurements are taken at the prescribed frequency, whether there is a clear assignment of responsibility for taking the measurements, and whether there are sufficient instructions. Auditors will also review the data and will verify that the data is indeed checked for compliance and, when appropriate, analyzed for trends.

The focus of ISO 14001 certification audit is the functioning of the environmental management system rather than compliance status. While auditors will review a sample of the compliance evaluation data, it is not going to be a systematic and thorough review. They will concentrate on verifying that there are internal procedures, resources and commitment to regularly evaluate compliance, and that the system works effectively. When auditors note a noncompliance against a regulation, they will be more concerned with the fact that the internal system did not identify the noncompliance, than with the actual noncompliance itself.

REQUIREMENT

Calibrate monitoring and measuring equipment, and maintain calibration records.

The purpose of this requirement is to ensure that the instruments and equipment used for taking measurements are suitable and accurate. Where instruments are not regularly calibrated and/or there are no documented calibration records, there can be no confidence in the measurement and the resulting data.

Companies operating a quality system, or even just inspecting products, are quite familiar with systems for controlling measuring and test equipment. ISO 9000 and most other quality management standards have detailed requirements for such calibration control system that far exceed the requirements of ISO 14001. If a company already operates a calibration control system for instruments and equipment used in inspection of products, there is no reason to establish a new, parallel system for controlling equipment used in the environmental system. It is sufficient to extend the scope of the existing calibration procedures to include equipment used for measuring environmental characteristics.

In those companies where there is no experience with a calibration control system, a new procedure must be established. At a minimum the measuring equipment must be uniquely identified, and be periodically checked and calibrated using approved calibration standards. The calibration should be recorded in a calibration certificate or a log. While most manufacturers and laboratories often calibrate their instruments in-house, companies that do not have previous experience with calibration are advised to send their instruments and equipment to an outside calibration service.

DOCUMENTATION

On the documentation level, companies that already have calibration control procedures need only reference these procedures in the environmental system manual, and modify the scope of the procedures to include instruments and equipment used for measuring environmental characteristics.

Where there is no such procedure, one must be established. At a minimum, the procedure should require that for each relevant gage, instrument or other piece of measuring equipment there be a calibration record, and that the equipment be marked with a sticker, or other means, indicating the next calibration due date. There should also be a calibration log or matrix to ensure effective tracking of equipment and its calibration status. Companies that calibrate their equipment in-house should also have instructions on how to calibrate various types of instruments and equipment, and how to establish calibration records. Standards used for calibration must be certified.

When equipment must be adjusted or calibrated before each use, there is no need to keep a record of such adjustments. But there must be a clearly posted warning, and instructions on how to adjust the equipment. If standard solutions or other types of reference standards are used, these must be certified or be prepared under controlled conditions.

EVIDENCE/RECORDS

The records and evidence of conformance are calibration control procedures and calibration records. Additional evidence of control is provided by a list or a log of equipment and its calibration status, and stickers identifying equipment with the next due date for calibration.

AUDIT

Verifying conformance, auditors will check the calibration status of instruments and equipment used in labs and installed on pipes, tanks, stacks and other points where environmental characteristics are measured; and will ask for corresponding calibration certificates. When calibration is performed in-house, auditors will also review calibration instructions and may ask for checking or recalibration of some instruments to verify that they were indeed accurately calibrated, and that personnel performing calibrations know how to use calibration procedures.

4.5.1 Nonconformance and corrective and preventive action

REQUIREMENT

Identify and investigate nonconformances and when appropriate take action to mitigate impacts.

Nonconformances are operations, activities, processes, products, and incidents that do not conform with specified requirements. The requirements may be defined in policies, objectives, targets, procedures, instructions, legislation, regulations, contracts, and in the ISO 14001 standard itself.

Several elements of the environmental management system are dedicated to identifying nonconformances. Examples include internal auditing and environmental performance monitoring system. In addition, nonconformances may be identified by external audits, regulators, customers, interested parties, and by anyone outside or within the organization who for any reason is aware of a nonconforming condition.

The system for handling environmental nonconformances should provide the means for documenting the nonconforming condition, for communicating the information to those responsible for mitigating possible impacts, for investigating the cause of the nonconformance, and for initiating a corrective action to prevent recurrence.

A specially designed form is often used to document, communicate and process nonconformances. At the top of the form there is a block for identifying date, time, location, event, originator, responsible organization, and other such particulars. Below there is a block reserved for describing the nonconforming condition. The next block is for establishing a record of impacts and any mitigating actions that may have been taken. And the last block is for reporting the causes of the nonconformance. At the bottom there is a space for noting whether a corrective action was requested or not. Corrective actions are initiated and processed using a different form (refer to the next requirement).

DOCUMENTATION

The standard requires that there be a procedure for handling nonconformances. Although there is no explicit requirement that the procedure be documented, it is difficult to imagine how it could be effectively communicated and implemented without being written down, especially in larger organizations.

The procedure should clearly define the responsibility and authority for handling and investigating nonconformances. There should also be a clear assignment of authority and some guidelines for taking mitigating actions. Different types of nonconformances may be handled by different functions and/or on a different level. For example, whereas nonconformances identified in the course of an internal audit are usually documented and followed up by the auditors, handling of a noncompliance against regulatory requirements may be the responsibility of the regulatory affairs staff.

The procedure should also instruct in the process of identifying a nonconformance, documenting it, communicating the information to relevant functions, and investigating causes. If a form, as described above, is used for this purpose, the procedure should instruct how to use and process the form.

EVIDENCE/RECORDS

The records and evidence of conformance are procedures, including formal assignments of responsibility and authority for handling nonconformances, and records of identified nonconformances and the mitigating and corrective actions associated with them. In some systems there may also be a log for tracking nonconformances and their status, and reports with nonconformance statistics.

AUDIT

Auditors will verify that there is a clear assignment of responsibility and authority for handling all types of nonconformances, and that the system is sufficiently open and flexible to identify and process all relevant nonconformances irrespective of their nature and origin. For example, auditors will object if the system is only used by functions responsible for monitoring environmental performance and regulatory compliance, but is not available to maintenance or customer service functions. It does not mean that all functions must use the same centralized system. Each function can have its own, if this is more natural and effective, but each individual system must meet the requirements of the standard and be formally incorporated into the environmental management system. Internal auditing almost always has its own independent system for reporting and handling nonconformances.

Auditors will interview personnel and review nonconformance files to verify that nonconformances are documented and are communicated to designated functions in a timely manner; that mitigating actions are taken when appropriate; and that causes of nonconformances are investigated in each case. They also will be interested in how the system for handling nonconformances interfaces with the system for initiating corrective and preventive actions (refer to the next requirement).

REQUIREMENT

Initiate and implement corrective actions to prevent recurrence of nonconformances.

The system for initiating and implementing corrective actions picks up where the system for handling nonconformances stops. Once a nonconformance is identified, documented, investigated and dealt with on the level of mitigating the immediate impacts, the corrective action system is responsible for dealing with the underlying causes to prevent recurrence of the same or similar nonconformance.

Not all nonconformances have to be automatically followed up with corrective actions, but all should be investigated to determine causes, and all should be evaluated to determine whether a corrective action should be initiated or not. The decision whether a corrective action will be initiated should be made by an authorized function and be based on some objective criteria. The decision must be recorded, so it is clear that a corrective action was not initiated because of a deliberate decision and not an oversight.

The corrective action system should be flexible to deal with very diverse types of actions, ranging from a minor correction of a procedure to a long-term project that may involve product design changes or introduction of new technology or equipment. The scope of a corrective action should be appropriate to the nature of the problem and the actual and potential environmental impacts.

Basic elements of the corrective action system are identification of the problem and causes; request for a corrective action; review and approval of a proposed action; implementation of the action; and verification that the action was implemented and that it is effective.

Corrective action request should be addressed directly to the department or function that is responsible for the nonconforming condition and is capable to address its causes. The responsible party should propose, in writing, the action to be taken and the estimated date of completion. The requesting party should then review and approve the proposed action and, on or immediately after the agreed completion date, should follow up to verify that the action has been implemented and that it is effective. The whole process can be documented using a corrective action request form. Separate blocks in the form are dedicated for the description of the nonconforming condition and its causes, proposal of the corrective action to be taken, and follow-up verification.

To help manage processing corrective actions it is useful to have a log for tracking their status. With many actions being processed simultaneously, and some taking a long time to implement, there is a real need for a mechanism to ensure that all corrective actions are completed and verified on time. A computer database or a spreadsheet file can be adapted for this purpose.

DOCUMENTATION

The operational procedure dealing with corrective actions should clearly define the function(s) vested with the authority to request corrective actions; explain the process; and instruct how to use the corrective action request form. When different functions and/or departments operate their own systems, each should have its own procedure.

EVIDENCE/RECORDS

Records and evidence of conformance are the corrective action requests and records of their implementation and verification. In addition, logs or other tracking systems provide the evidence that processing of corrective actions is adequately managed.

AUDIT

Auditors will review the corrective action procedure to verify that the responsibility and authority for requesting corrective actions and for checking their effectiveness is clearly defined, and that there are adequate instructions for using the system. They will also verify that all relevant nonconformances are followed up with requests for corrective actions, and that corrective actions are processed, implemented and closed out in a timely manner. In larger organizations, where several corrective actions may be processed at the same time, auditors will expect a log for tracking the status of opened corrective actions.

REQUIREMENT

Initiate and implement preventive actions to prevent occurrence of potential nonconformances.

There is often some confusion about the difference between corrective and preventive action. Many people incorrectly believe that preventive action is a part of corrective action — the part that prevents a nonconformance from happening again. While it is true that a corrective action must have an element preventing recurrence of the same or similar nonconformance, in ISO 14001 a preventive action is taken to prevent occurrence of a potential nonconformance, and is distinct from a corrective action that is taken to address an actual nonconformance. In practice, corrective actions are implemented following an actual spill, release, or other actual event; while preventive actions are implemented when nothing has yet happened, but there is a pattern of decreasing control or capability that needs to be addressed before it is too late.

To have a meaningful preventive action system there should be a systematic effort to identify potential problems. The usual methods are inspections, monitoring, statistical process control, and review of records and data to identify patterns of decreasing control or capability. Once the potential problems are identified, processing of preventive actions can follow the same procedure that applies to corrective actions. The difference is only in the first step of identifying the problem.

DOCUMENTATION

Corrective and preventive action systems are usually documented in the same procedure. An introductory section in the procedure explains the difference between the two types of actions, and there are separate sections for initiating corrective and preventive actions. For preventive actions the procedure should explain how potential problems are identified and who authorizes and/or requests preventive actions. From there on the procedure is usually identical, prescribing the same forms, records, processing, and closing methods for both types of actions.

EVIDENCE/RECORDS

Records and evidence of conformance are the preventive action requests (using the same forms as for corrective actions), and records of their implementation and verification. In addition, there should be some evidence that there are programs of regular inspections and/or monitoring to identify potential problems and nonconformances; for example, periodical inspections of tanks, testing of alarm systems, statistical process control, and so forth.

AUDIT

Assessing the preventive action system, auditors will verify that there are established methods for identifying potential problems, that the identified potential problems are being addressed using formal requests for preventive actions, that preventive actions are implemented in a timely manner, and that their implementation is verified.

In practice, the most frequent problem is that companies do not understand the distinction between corrective and preventive actions, and therefore do not have any formal systems for identifying potential problems and for issuing preventive action requests. Often, auditors find evidence of inspections, audits and other systematic efforts to identify potential problems, and even evidence that preventive actions are taken, but these activities are not formally organized into a preventive action system and are not defined in the environmental management system.

4.5.3 Records

REQUIREMENT

Maintain records necessary to demonstrate conformance to the requirements of ISO 14001.

While the first paragraph of Section 4.5.3 explicitly names training, audit and review records, the scope of records that must be maintained is really defined in the last paragraph of this section: "Records shall be maintained . . . to demonstrate conformance to the requirements of this international standard."

It is indeed a very broad scope, including records of all activities that are required by ISO 14001. In practice, this requirement allows auditors to ask for written records, reports, minutes of meetings, etc., documenting results of such diverse activities as selection of significant aspects, training, emergency drills, compliance evaluation audits, corrective actions, management reviews, and so forth. In fact, this requirement creates an obligation for the audited organization to provide written evidence of conformance with all and any requirement of ISO 14001.

DOCUMENTATION

The scope of records maintained by an organization is usually documented in procedures dealing with activities that require establishment of records. For example, a procedure dealing with management reviews would define what kinds of records must be established to document results and conclusions of the reviews, and instruct how to establish these records. There is no need for a global list of all records to satisfy this requirement. However, companies often end up compiling such a list to document retention times of records and locations where the records are kept (see last requirement of this section).

EVIDENCE/RECORDS

The evidence of conformance is the actual availability of all records necessary to demonstrate conformance with the standard.

AUDIT

Nonconformance with this requirement, i.e., lack of records, makes it impossible for auditors to obtain objective evidence that the audited activity is carried out in accordance with planned arrangements (procedures), and even that it has been at all implemented. In some cases auditors can obtain the evidence by asking questions and by observation. However, most activities required by the standard would not be carried out during the audit, and observation would not provide any information with regard to past performance. Existence of relevant records is therefore crucial to the very feasibility of conducting the certification audit. Lack of objective evidence of conformance will result in interruption of the audit process, or a major nonconformance, depending on the circumstances.

REQUIREMENT

Index and organize environmental records to facilitate their retrieval, and protect records against damage or loss.

Section 4.5.3 mentions identification, traceability, storage, maintenance and disposition of environmental records. All these words really mean one thing: retrievability. Any record management system will be accepted as long as it works. When a record has to be retrieved, it should be easily found and be legible and in good condition.

Records should be filed and stored in designated locations. The cabinets, drawers, shelves, and files containing records should be clearly labeled. Records should be arranged and indexed by product, process or area that they concern, by event or project to which they pertain, or by date. When appropriate, the indexing should be cross-referenced. Specific function (person) should be assigned with the responsibility for maintaining record files, including records in electronic format.

DOCUMENTATION

The operational procedure dealing with control of environmental records should define the requirements for indexing, filing, and maintaining records, and should assign the responsibility for these activities. Preservation and retrievability of records should be the ultimate objective of the record management system.

EVIDENCE/RECORDS

The evidence of conformance consists of record management procedures and the actual condition, organization and retrievability of records.

AUDIT

The functioning of the environmental record management system is continuously tested throughout the whole audit. Auditors will ask for retrieval of records when assessing almost any activity of the environmental system. Records are often the only evidence that certain activities are taking place. (For example: management reviews, internal audits, corrective actions, training, and many others.)

When auditors ask for a record, they will note how long it takes to retrieve it. If it happens that a record cannot be found immediately, it is best to tell the auditor right away that there will be some delay in retrieving the record, and encourage him or her to continue with the audit in the meantime. Nothing is more embarrassing and counter-productive than having the auditor wait and watch managers and staff frantically searching for a record.

When a record cannot be located at all, auditors will usually investigate the reason. If the problem is the lack of proper indexing and organization of records, there may be two nonconformances: one for the audited activity (lack of evidence of conformance) and the other for the breakdown of the record management system.

REQUIREMENT

Determine and document retention times for environmental records.

ISO 14001 does not specify any particular retention times for records, but it requires that organizations themselves decide how long they will keep their environmental records, and that they document the decision.

Retention times should be determined individually for each type of record, based on commonsense considerations of their useful life. For example, records pertaining to periodical assessments, reviews, and certifications, such as internal audit reports, management review records, or calibration certificates, are usually kept for three cycles; i.e., records of annual events are kept for three years. Retention time for training records can be established on the basis of employment status. Retention time for some records will be dictated by legal, regulatory or contractual requirements.

DOCUMENTATION

The procedure dealing with the control of environmental records should list all the categories of records and stipulate for each the retention time. As an alternative, retention times may be specified directly in procedures dealing with activities that require establishment of records. The procedures would instruct how to establish the record, where and how it should be stored, and how long it should be kept.

For categories of records whose disposition depends on an event rather than just passage of time, the retention period can be defined in terms of the event instead of absolute time. For example, it can be stipulated that training records are retained for two years after termination of employment.

EVIDENCE/RECORDS The evidence of conformance consists of procedures documenting retention times for records, and the actual implementation of the procedures.

AUDIT Assessing conformance, auditors will review the procedure stipulating retention times for records, to ensure that all relevant records are included and that the specified retention times make sense. Auditors will usually accept the specified retention times unless there is an obvious problem with a record being discarded too quickly, or when there is a conflict with legal or regulatory requirements. To test implementation, auditors will ask for retrieval of some older records from various categories.

4.5.4 Environmental management system audit

For those readers who are not fully familiar with internal auditing there is a short description of the audit process included in Chapter 3, *Definitions*, under the definition of *Environmental Management System Audit*. The definition and its explanation should help to better understand the intent of the following four requirements under this section.

REQUIREMENT

Establish a program, including scope and frequency, for conducting environmental management system audits.

Audits required in this section may be conducted by personnel from within the organization or by external persons (consultants) contracted by the organization. These types of audits are known as internal audits.

All activities comprising the environmental management system should be audited at least once a year. Systems under implementation, and for the first year or two of operation, should be audited more frequently: every three or six months, for example.

Even for mature systems, those activities that are especially important, or where there are chronic implementation problems, should be audited more frequently. The standard explicitly requires that the audit program be based on the importance of activities and results of previous audits.

The audit can be carried out all at once, as certification audits are, or be distributed throughout the whole auditing cycle. The distributed way is preferable in most cases.

A popular format for an audit plan is a matrix with vertical listings of all environmental system elements (activities) and horizontal listings of all relevant departments and areas in the company. In the blocks where the elements and areas intersect, the planned audit dates can be written in or, if a given activity is not relevant in an area, the block can be crossed out. This format ensures that all environmental system elements will be audited in all relevant areas; i.e., that nothing will be missed. Certification auditors use such matrices when planning their audits.

DOCUMENTATION

The procedure dealing with internal audits should explain what criteria need to be applied when planning internal audits; how to establish and document the audit plan; and who is responsible for establishing the plan and monitoring its execution.

EVIDENCE/RECORDS

The record and evidence of conformance of this activity is the internal audit plan itself.

AUDIT

Auditors will review the internal audit plan, noting if all relevant activities and areas are included, and if the frequency of auditing reflects the maturity of the environmental system and the importance of individual elements. Auditors will not accept audit plans where each activity and area is scheduled for only one audit during the cycle, without any regard for importance and/or performance of individual activities.

REQUIREMENT

Define the methodology and requirements for conducting environmental management system audits.

ISO 14001 does not contain any specific requirements for conducting internal audits. The standard only states the objectives for the audit and requires that there be a comprehensive procedure defining scope, frequency, methodology, responsibilities, and requirements for conducting internal audits and reporting results. But it does not contain any specific criteria with regard to auditor qualifications, audit methodology or reporting. Thus, in theory at least, organizations may define their requirements as they see fit.

In practice, certification auditors will expect that the internal audit system meets certain generally accepted criteria, even though these are not spelled out in the standard. The generally accepted criteria include:

- Defined auditor qualification requirements;
- Consideration of auditor independence;
- Written instructions for preparing and conducting the audit;
- Defined requirements for reporting audit results; and
- Provisions for follow-up audits to verify that corrective actions have been implemented and that they are effective.

Qualifications of auditors should be evidenced by training records. These can be certificates from courses and seminars for auditors, or internal training provided by a consultant or someone in the company who is already qualified. Self-study training can also be accepted, but it must be formally recorded.

Whenever possible, auditors should be independent from those responsible for the audited activity. Exception can be made when only one person in the organization has the necessary expertise in the audited area. Otherwise, certification auditors will object to self-audits.

Auditors should prepare for each audit. The preparation may comprise review of the ISO 14001 sections and environmental system procedures relevant to the audited activity and review of environmental records, including audit reports from previous audits, nonconformance reports, and corrective action requests. It is recommended that auditors use checklists. Preparing for the audit, auditors should establish checklists, or supplement standard checklists, to include questions pertaining to current projects and issues. Internal auditors should focus on specific current and potential problems, rather than mechanically ticking off a standard checklist.

DOCUMENTATION

The procedure for internal auditing should specify qualification and training requirements for auditors and instruct how to prepare for and conduct the audit. The procedure should also express the policy to use auditors independent from the audited activity whenever possible. The internal audit procedure must be detailed and comprehensive.

EVIDENCE/RECORDS

The records and evidence of conformance are the auditor training and qualification records, and audit checklists, if these are retained. Also, the audit reports themselves will provide the evidence of implementation.

AUDIT

Assessing conformance, certification auditors will interview some internal auditors, questioning how they learned to conduct audits and how they prepare for the audit. Reviewing internal audit reports, auditors will also note if the assigned internal auditors are independent of the audited activity. Although auditor independence is not explicitly required in ISO 14001, it is a generally recognized principle.

REQUIREMENT

Implement corrective actions to deal with nonconformances identified by the audit.

Every nonconformance identified during an internal audit should automatically cause a request for corrective action. While in other circumstances the need for a corrective action can be evaluated and decided upon, there is no such freedom in the case of internal audits, especially when the nonconformance is against requirements of the environmental management system. In order to maintain certification, the system must comply with ISO 14001 and other planned arrangements at all times, without exception.

For this reason, the internal audits, corrective action requests, and follow-up audits are usually integrated into one system. The forms used for reporting internal audits are adopted for processing all three phases.

But this integrated processing is the only difference. Otherwise, corrective actions resulting from internal audits must follow exactly the same rules that apply in general to corrective actions (see Section 4.5.2).

DOCUMENTATION

The operational procedure dealing with internal audits should clearly state the policy that all identified deficiencies must be followed up by corrective actions. The procedure should explain the process and responsibilities for requesting, implementing, and verifying corrective actions, or refer to another procedure that already explains the corrective action initiation and implementation process (see Section 4.5.2).

EVIDENCE/RECORDS

The records and evidence of conformance are the internal audit reports, in particular the sections reporting on corrective actions and follow-up audits.

AUDIT

Assessing conformance, auditors will review the internal audit reports, noting if corrective actions are being implemented in a timely manner, and if all are followed up to verify their implementation and effectiveness.

REQUIREMENT

Report the results of audits to the executive management.

A popular system for reporting internal audits is based on a one-page form for recording and processing each individual nonconformance. The form is very similar to that used for processing and recording corrective actions. It has the following four blocks. *Heading*: for recording the particulars of the audit, such as identification of the audited location and activity, names of the auditor and the manager responsible for the audited area, etc. *Nonconformance:* for describing the nonconforming condition and classification of the nonconformance. *Corrective Action*: for describing the proposed action and the agreed implementation due date. *Follow up*: for recording the result of the follow-up audit and closing the nonconformance.

When a nonconformance is noted, the auditor fills out the *Heading* and the *Nonconformance* blocks, and passes the form to the manager responsible for the audited area, who uses the third block, *Corrective Action*, to propose a corrective action. On or shortly after the implementation due date, the auditor comes back to the area to verify that the corrective action has been implemented and that it is effective. The result of the follow-up audit is recorded in the last block of the form, *Follow up*. The final report for the audit can be put together by adding a cover page to the individual nonconformance reports. The cover page should provide audit particulars (date of the audit, audited areas, names of auditors, etc.) and a brief summary of audit results.

At the end of the auditing cycle, when all activities and areas have been audited, the partial audit reports should be bound together and submitted for review to the executive management. This is normally done in conjunction with the scheduled management reviews of the environmental system (see Section 4.6). To facilitate the review, the final report should also contain statistics, conclusions, and recommendations.

DOCUMENTATION

The operational procedure dealing with internal audits should explain how to use the audit nonconformance report forms; instruct how to prepare the final report for the whole audit cycle; assign the responsibility for preparing the report; and explain how the report should be submitted and/or presented for the management review.

EVIDENCE/RECORDS

The records are, of course, the internal audit reports themselves.

AUDIT

Assessing conformance, auditors will review the internal audit reports, noting if they are established, processed, and evaluated by management in accordance with the governing procedures. A common problem with nonconformance reports is that corrective actions are not followed up in a timely manner.

4.6 MANAGEMENT REVIEW

4.6 MANAGEMENT REVIEW

This section defines requirements for the last element of the environmental management system cycle. The first step in the cycle was formulating the environmental policy; the second and third steps were planning and implementing a system for achieving the policy; the fourth step was monitoring and checking the system and its performance; and this last step is a top-level review to assess whether the system is suitable, adequate and effective, and to identify opportunities for improvement.

The management review should be concluded with specific recommendations for changes to address new developments and to improve the environmental management system. The changes should then be implemented at each step of the next management system cycle, i.e., be incorporated into the policy, planned for and implemented into the system, monitored and checked for effectiveness; and finally reviewed by the next management review.

Management reviews have thus the role of launching the environmental management system into the next cycle of improving the system and its effectiveness.

REQUIREMENT

Periodically conduct management reviews of the environmental system to ensure its continuing suitability, adequacy and effectiveness.

Management reviews should be conducted by the executive management with authority to formulate policies and allocate resources for the environmental management system. The reviews should not be confused with operational level meetings for coordinating and monitoring implementation and operation of the system. While operational issues and data may be discussed to provide baseline information about performance of the system, the main purpose of management reviews is a global assessment of continuing suitability and adequacy of the environmental policy and objectives, and the effectiveness of the environmental management system.

Management reviews must be conducted regularly. Quarterly or half-yearly reviews are recommended for new systems under implementation and maturation. Mature systems can be reviewed on an annual basis. In addition to scheduled reviews there should also be allowance for calling extraordinary review meetings in response to sudden important developments or emergency situations. If possible, management reviews should be synchronized with the cycle and timing of internal audits, so that the final audit report is available for the review.

The most common format for the management reviews is a meeting; but other methods, such as telecommunication conferences, or circulation of reports, memoranda, etc., are also acceptable. The participants should include the top executive management and managers of such functions as marketing, engineering, production and, of course, functions responsible for day-to-day operation of the environmental management system.

DOCUMENTATION

A dedicated operational procedure should be established to set out specific instructions for conducting management reviews. The procedure should define the frequency, participation, and scope for the reviews, and explain how to schedule, conduct, and record the reviews. Smaller companies can dispense with the procedure and document everything in the environmental system manual.

EVIDENCE/RECORDS

Records of management reviews are discussed in the last requirement of this section.

AUDIT

Auditors will spend considerable time evaluating management reviews. Proper functioning of this element of the environmental system is considered by registrars and third-party auditors to be crucial to the success of the whole system. As was the case with the environmental policy, it is likely that the whole audit team will participate in auditing management reviews. Auditors will interview the management personnel, examine the management review records, and evaluate other evidence to verify that management reviews are conducted regularly as scheduled; that the agenda for the reviews is relevant and appropriate; that review records are maintained; that conclusions and recommendations of the reviews are being implemented; and that, overall, management reviews fulfill their role of launching the environmental system into the new cycle of continual improvement.

REQUIREMENT

Provide management with information necessary to carry out the review.

ISO 14001 explicitly requires that the executive management be provided with the information necessary to conduct a meaningful review. At a minimum, this information should include:

- Reports on the progress toward achieving the environmental policy, objectives and targets (including those related to the prevention of pollution);
- Status of compliance with environmental laws and regulations;
- Environmental performance in areas related to significant aspects;
- Results of internal audits of the environmental management system;
- Achievements and status of continual improvement projects; and
- Information about any developments and changes in products, processes, operations, laws, regulations, views of interested parties, etc., that may create new environmental aspects or impact the significance rating of previously identified aspects.

In addition, although not required by the standard, the executive management could benefit from such information as the cost of operating the environmental system; savings attributable to the system (for example, savings in hazardous waste disposal cost); benchmarking information comparing the organization's environmental performance to others within and outside the industry; public and market response to the organization's environmental initiatives; and so forth.

Information required by the management may be provided in any form. It may be contained in reports, memoranda, summaries, etc., or be presented directly at the management review meeting. For example, during the meeting, various departmental managers, committees, groups, etc., may be asked to present data and results in their areas of responsibility or competence.

Documentation

The environmental system manual and the operational procedure for conducting management reviews should require that pertinent information bc collccted and provided to the executive management for the review. In larger organizations the procedure should also document how the information is collected and presented, and who is responsible for doing it. In a small company with a few significant aspects and minor regulatory issues, it is sufficient to state a commitment to provide the information and outline its scope.

Evidence/Records

The records and evidence of conformance consist of reports, summaries, charts, and other such documents providing the executive management with baseline information for their review of the environmental management system. If oral presentations are made during the management meeting itself, they should be recorded in the minutes of the meeting.

Audit

Auditors will ask directly what information was provided to the executive management for their review of the environmental system, and will want to see the pertinent documents and records. Auditors will expect the scope of information to be at least that corresponding to the bulleted list included in the first paragraph of this section.

REQUIREMENT

Ensure that management review considers the possible need for changes to policy, objectives and other elements of the environmental system.

The principal role of management reviews is to ensure continuing suitability and effectiveness of the environmental system, and to launch the system into the next cycle of improvement. This philosophy is at the core of the management system model on which ISO 14001 is based.

In line with this role, there is an implied expectation that management reviews should be concluded with recommendations for changes to the policy, objectives or other elements of the system. Especially in the implementation and maturation phase, failure of a management review to recommend specific changes implies a general lack of commitment to the environmental system, or a lack of understanding of the principles on which the system is based. Even in mature systems the changing circumstances and commitment to continual improvement should compel the executive management to adjust the environmental objectives and set new goals for further improvement of the system.

DOCUMENTATION

On the documentation level, the environmental system manual should express the importance of management reviews actively considering the need for changes to policy, objectives and the environmental system. The procedure for conducting management reviews may further explain and reinforce this policy.

EVIDENCE/RECORDS

The record and evidence of conformance are reports, summaries, or minutes of meetings documenting recommendations for changes to the policy, objectives or other elements of the environmental system. If a management review is concluded without any recommendations for changes, there should be at least evidence that the possible need for changes was considered.

AUDIT

The ISO 14001 standard does not explicitly require every management review to be concluded with recommendations for specific changes, and a failure to do so is not a nonconformance. However, to auditors, a management review concluding without any recommendations for changes is a warning that there may be a lack of commitment to maintain and improve the environmental system, or that the executive management does not understand the role and purpose of the review. Suspecting a serious breakdown in the system, auditors will thoroughly examine all the facts associated with the review in question, and will try to develop their own independent opinion whether changes should have been recommended or not.

REQUIREMENT

Ensure that the review addresses the commitment to continual improvement.

ISO 14001 requirements for continual improvement are weak, confusing, and even contradictory. There are two problems. One is that the requirement is only for improving the environmental management system, and not environmental performance; and the other is that there are no planning, operational and checking requirements related to the continual improvement.

By requiring in the environmental policy a commitment to continual improvement, the standard sets out continual improvement as a very important element of the system — comparable to prevention of pollution and commitment to comply with laws and regulations. However, after such a formidable introduction the issue is dropped. The next time continual improvement is referred to is in the last sentence of the *Management Review* section, and even there it is just mentioned.

The failure of the standard to define any specific requirements with regard to the system for managing continual improvement should not, however, be interpreted as a lack of requirement for demonstrating continual improvement as such. Once a commitment for continual improvement is stated in the environmental policy — and it is required — the organization must be able to demonstrate that it has implemented the commitment.

Any program or project that aims at improving the effectiveness of the environmental management system and/or environmental performance can be used as evidence to demonstrate implementation of the commitment for continual improvement. On the system level, improvement can be demonstrated by changes in procedures; establishment of new instructions; additional training; and other such changes and activities improving effectiveness of the system. On the environmental performance level, it is the data demonstrating attainment of new objectives and targets, and improved compliance.

DOCUMENTATION

The environmental system manual should explain the meaning of the commitment to continual improvement and what kinds of programs, projects and activities should be established to implement this commitment. The manual should also define the role of management reviews in evaluating whether the commitment to continual improvement is satisfactorily implemented. The procedure for conducting management reviews may further require that continual improvement be always included in the reviews, and that conclusions and recommendations regarding continual improvement be always recorded.

EVIDENCE/RECORDS

The evidence of conformance consists of documents and records pertaining to continual improvement programs and projects, and agendas and records of management reviews. The record must demonstrate that the management actively reviews the implementation status of the organization's commitment to continual improvement.

AUDIT

Auditors will be looking for evidence that the management considers its commitment to continual improvement as a reference in reviewing the environmental management system. They will interview management personnel to verify that there is a clear understanding in the organization how the commitment to continual improvement should be implemented, and that the managers can point out specific programs, projects and activities for implementing this commitment. Auditors will also examine the management review records to verify that continual improvement is considered at every review, and that, when appropriate, there are some pertinent conclusions or recommendations.

REQUIREMENT

Document the management review and record its results.

ISO 14001 explicitly requires that management reviews be documented. The nature of the documentation and records will depend on the format of the review itself. It can be agendas, attendance lists, presentations, reports, minutes of meetings, memoranda, action lists, and so forth. The documents and records should identify the date, place, attendance, and agenda of the review, and should provide at least a summary of the review and its conclusions and recommendations.

There may be some important legal aspects associated with management reviews. They are discussed in *ISO 14000 and the Law* by S. Wayne Rosenbaum.

DOCUMENTATION

The operational procedure for conducting management reviews should instruct how to document the review and record its conclusions and recommendations. The procedure should also specify the initial distribution of the documents and the storage location and retention period for the records.

EVIDENCE/RECORDS

The evidence of conformance are the actual management review documents and records.

AUDIT

Auditors never have the chance to witness or participate in management reviews. Documents and records are the primary evidence of conformance. It is, therefore, especially important that the records are properly established and maintained. Auditors will check that appropriate documents and records are maintained and that their content is sufficient to determine the attendance, agenda, recommendations and conclusions of the review. Records must be sufficiently detailed to allow auditors to verify conformance with all requirements of this section. For example, a verbal statement that continual improvement was considered at the review is not sufficient. The minutes of the review meeting, an action list, or some other document must demonstrate that it was indeed considered.

DOCUMENTATION REQUIREMENTS FOR THE CERTIFICATION AUDIT

The requirements are liberal ....

ISO 14001 requirements with regard to the documentation of the environmental management system are very liberal. In most sections the standard requires that "The organization shall establish and maintain procedures to . . .", but only a few sections explicitly require these procedures to be documented. The intention was to avoid bureaucratic and rigid systems, such as ISO 9000.

While this intention is clear and should be respected, in real life, those organizations that wish to have their environmental management systems certified by an independent registrar must document the whole system in procedures, whether such procedures are explicitly required or not. There is simply no known method to audit a management system that is not documented.

... But auditors will find ways to force organizations to completely document their environmental systems.

ISO 14001 auditors will not be able to write findings for lack of documentation. However, where they feel that a written procedure should be established, they will find ways to force the issue. For example, auditors can interview personnel to verify that everyone in the organization has the same knowledge and understanding of what the unwritten procedures are. If there are any discrepancies, they will issue findings stating that personnel are not familiar with procedures, which will in practice force the organization to document its system. There is also Section 4.4.6 a) requiring that organizations shall "establish and maintain documented procedures where their absence could lead to deviations form the environmental policy and the objectives and targets". Another place where auditors can rest their finger is Section 4.4.4 requiring that "The organization shall establish and maintain information, in paper or electronic form, to a) describe the core elements of the management system and their interaction; b) provide direction to related documentation".

Thus, the intention of the framers of the standard to make written procedures voluntary will not hold in a certification audit. Auditors have a legitimate need to have the system documented and have sufficient support in the standard itself to require, whether directly or indirectly, complete documentation.

The system is usually documented on four levels:

Environmental management system documentation is usually structured on four levels. The top-level document is the environmental manual. The second level are the operational procedures. The third level are programs for achieving objectives and targets. And the fourth level are work instructions, specifications, standards, data and other such documents.

Envrionmental Manual

The role of the environmental manual is to state the environmental policy; to outline the structure and to summarize the major procedures of the system; to define the principal responsibilities and authorities; and to reference other documents supporting the system, especially the operational procedures.

Operational Procedures

Operational procedures define individual elements of the management system. They describe the pertinent activities (what and when), designate the responsible personnel (who), and provide system-level instructions (how). The distinction between operational procedures and work instructions is that procedures explain systems and activities, while work instructions provide specific steps for accomplishing a task.

Management Programs

Documents defining the management programs for achieving objectives and targets define specific goals and milestones that must be reached within certain time frames, specify the means and methods to be used for reaching the goals, assign personnel for managing and carrying out the improvement projects, specify how progress toward achieving the goals should be measured, and prescribe requirements for reporting the results.

Work Instructions

Work instructions are step-by-step instructions on how to carry out a specific task, for example: how to label and handle hazardous materials, how to clean up a small spill, how to use protective equipment, how to evacuate an area, how to take a measurement, etc.

Sample of a generic environmental manual and operational procedures are provided in *ISO 14001 Documentation* (annotated on the next page).

ISO 13485 & EN 46000

by Lynette Howard and Jack Kanholm

ISO 13485 (EN 46000) QUALITY SYSTEM MANUAL AND 36 PROCEDURES

Book and computer software with quality manual, procedures and forms - $ 390.

ISO 13485 (EN 46000) REQUIREMENTS

99 Requirements Checklist and Compliance Guide

170 Pages
Hardcover Book - $ 59.

ISO 13485 (EN 46000) IN OUR COMPANY

Self-Study Course for Personnel

32-Page Booklet, Test and Certificate - $ 9.

QUALITY SYSTEM FOR THE MEDICAL DEVICE INDUSTRY

The three publications in this series provide: ready-made template documentation with the quality manual and operational procedures, explanation of all ISO 13485 and EN 46000 requirements, and a general orientation course for personnel. All three publications are based on Jack Kanholm's ISO 9000 series (refer to the preceding page annotating the ISO 9000 publications).

Ms. Howard contributed her unique knowledge of both the U.S. and European regulatory requirements and certification systems.

QS-9000 by Jack Kanholm

QS-9000 QUALITY SYSTEM MANUAL AND 40 PROCEDURES

Book and computer software with quality manual, procedures and forms - $ 390.

QS-9000 REQUIREMENTS

107 Requirements Checklist and Compliance Guide

177 Pages
Hardcover Book - $ 59.

QS-9000 IN OUR COMPANY

Self-Study Course for Personnel

32-Page Booklet, Test and Certificate - $ 9.

QUALITY SYSTEM FOR THE AUTOMOTIVE INDUSTRY

The three publications in this series provide: ready-made template documentation with the quality manual and operational procedures, explanation of all QS-9000 requirements, and a QS-9000 general orientation course for personnel. All three publications are based on Jack Kanholm's ISO 9000 series (refer to the preceding page annotating the ISO 9000 publications).

Since 1995 thousands of companies have used these materials to successfully achieve QS-9000 certification.

More info on the internet at www.AQAco.com

ORDER FORM

Qty	Title	Prc	Dsc	Total	Qty	Title	Prc	Dsc	Total
	ISO 9000					**ISO 14000**			
	ISO 9000 Requirements	$ 49.	%			ISO 14001 Requirements	$ 49.	%	
	ISO 9000 Quality System	$ 69.	%			ISO 14001 And The Law	$ 59.	%	
	ISO 9000 In Our Company	$ 9.	%			ISO 14001 In Our Company	$ 9.	%	
	ISO 9000 Template Manual and Procedures Software	$ 290.	%			ISO 14001 Template Manual and Procedures Software	$ 290.	%	
	QS-9000					**ISO 13485 (EN 46000)**			
	QS-9000 Requirements	$ 59.	%			ISO 13485 Requirements	$ 59.	%	
	QS-9000 In Our Company	$ 9.	%			ISO 13485 In Our Company	$ 9.	%	
	QS-9000 Template Manual and Procedures Software	$ 390.	%			ISO 13485 Template Manual and Procedures Software	$ 390.	%	

Sales tax of 7.75% (CA only) and shipping cost (see chart below) will be added to invoice

Quantity Discounts (copies per title)

5 to 9: 10% 20 to 39: 30% Over 100: 50%
10 to 19: 20% 40 to 99: 40%

Shipping (by UPS)

❑ Ground $ 6. ❑ 2nd Day $ 13.
❑ 3rd Day $ 9. ❑ Next Day $ 25.

Shipping Address (No PO Boxes)

Mr. ❑ Ms. ❑

Title:

Company:

Street:

City: State: Zip:

Phone & Fax:

Billing Address

Company:

Street:

City: State: Zip:

Attention:

Method of Payment

Card No.:

❑ Check ❑ Visa ❑ MC ❑ AmEx Exp.:

❑ Bill Company Purchase Order No.:

Signature & Date:

30 Day Preview

❑ Yes, I want to preview the publications and reserve the right to return them for a refund (excluding shipping cost), if not satisfied.
I understand that to receive a refund, I must return the publications within 30 days.

To order, fax this form to (626) 796 9070 or call (800) 600 3601

ISO 14001 REQUIREMENTS